Cosmic Echoes

Searching For Ancient Life Among The Stars

GEW Reports & Analyses Team

Global East-West (London)

Global East-West (London).

CONTENTS

Chapter One

Introduction: Bridging Cosmology and Archaeology

The Intersection of Science and History

As we embark on an exploration of the intersection between cosmology and archaeology, it becomes evident that both fields are intricately intertwined by a shared quest for understanding humanity's origins and its place in the universe. Cosmology delves into the mysteries of the cosmos, seeking to unveil the fundamental principles governing the universe's birth, evolution, and ultimate fate. Archaeology, on the other hand, meticulously reconstructs the story of human civilization through the examination of material culture and historical artifacts. Despite their seemingly divergent scopes, these disciplines converge in profound ways, shedding light on the interconnectedness of cosmic events and human history. At the heart of this synergy lies the recognition that cosmic phenomena have significantly influenced the course of human civilizations, shaping belief systems, cultural practices, and societal developments. Moreover, archaeological findings often provide tangible evidence of ancient societies' responses to celestial occurrences, such as lunar calendars, solar alignments, and mythological representations of celestial bodies. This symbiotic relationship invites us to rethink traditional boundaries and acknowledge the profound impact of astronomical events on the collective human experience. By exploring this dynamic interplay, we not only gain a deeper understanding of our cosmic heritage but also reframe our interpretation of historical events within the broader context of the universe's grand narrative. In the chapters that follow, we will further unravel the intricate tapestry woven by the confluence of science and history, illuminating the enduring dialogue between cosmic phenomena and human endeavors.

Defining Cosmology and Archaeology

Cosmology and archaeology represent two distinct yet intricately interconnected fields of study that shed light on humanity's place in the universe. Cosmology, often referred to as the science of the universe, explores the origins, evolution, and eventual fate of the cosmos through a combination of theoretical frameworks, empirical evidence, and astronomical observations. This discipline delves into fundamental questions regarding the composition of the universe, the nature of dark matter and dark energy, the cosmic microwave background radiation, and the formation of galaxies and celestial bodies. On the other hand, archaeology is focused on uncovering the material remains of past human cultures, civilizations, and societies. By meticulously excavating and analyzing artifacts, structures, and human settlements, archaeologists reconstruct and interpret ancient lifeways, social organizations, and technological advancements. While seemingly disparate at first glance, these two fields converge in the realm of cosmic archaeology, where the traditional boundaries between terrestrial and extraterrestrial perspectives are transcended. The fusion of cosmological theories with archaeological methodologies opens up new avenues for unraveling the deep-time complexities of our cosmic heritage. By examining ancient artifacts and cultural traces from the perspective of cosmic evolution and potential extraterrestrial influences, researchers aim to bridge the gap between cosmic phenomena and human history.

Furthermore, the interdisciplinary nature of this endeavor necessitates a synergistic approach, wherein cosmologists, astronomers, anthropologists, and archaeologists collaborate to confront the challenges inherent in exploring the nexus of space and time. Through an exploration of the concept of cosmic archaeology, this chapter aims to elucidate the profound implications of integrating cosmological and archaeological paradigms, thereby fostering a deeper understanding of our cosmic origins and the potential interconnectedness of life across the universe.

Historical Perspectives on Cosmic Archaeology

Throughout history, humanity has been captivated by the mysteries of the cosmos and the origins of life on Earth. The quest to understand our place in the universe has driven both cosmologists and archaeologists to seek evidence that sheds light on the interconnectedness of celestial phenomena and terrestrial existence. In antiquity, various civilizations observed and documented celestial events, attributing profound symbolic and religious significance to the movements of the stars and planets. The ancient Egyptians, for instance, aligned their pyramids and temples with astronomical precision, emphasizing a deep connection between their earthly structures and the heavens above. Similarly, the Mayans meticulously recorded celestial observations, creating complex calendars and intricate cosmological frameworks that underscored the importance of cosmic influences on human affairs. The early pioneers of archaeoastronomy, such as Sir Norman Lockyer in the late 19th century, recognized the potential links between ancient monuments and astronomical phenomena. Their groundbreaking work laid the foundations for the interdisciplinary field of cosmic archaeology, which seeks to uncover the celestial knowledge embedded within archaeological remnants. As the 20th century unfolded, scholars increasingly delved into comparative studies of ancient mythologies, folklore, and oral traditions, pinpointing recurring motifs and narratives that reflect cosmological beliefs rooted in the distant past. Furthermore, historical texts and artifacts have provided valuable insights into how ancient cultures conceptualized the cosmos and its influence on their societal development. From the celestial observations recorded in the Enuma Elish of Mesopotamia to the astronomical alignments of Stonehenge in prehistoric Britain, these cultural remnants offer compelling evidence of humanity's enduring fascination with the celestial realm. As the tools of modern science continue to evolve, the interplay between cosmology and archaeology has entered a new era of discovery. Advancements in spectroscopy, radiocarbon dating, and space telescopes have enabled researchers to analyze ancient relics with unprecedented precision, revealing hidden astronomical symbolism and knowledge encoded within material culture. By combining data from cosmic observations with archaeological findings, scholars are unraveling how ancient civilizations navigated the cosmos, demonstrating the integral role of celestial phenomena in shaping human cultural landscapes. This historical contextualization serves as the basis for comprehending the profound implications of cosmic archaeology and its relevance in understanding both the past and the present.

Methodological Approaches in Cosmology and Archaeology

In the interdisciplinary field of Cosmic Archaeology, methodological approaches play a pivotal role in unraveling the mysteries of our cosmic origins. Incorporating cutting-edge techniques from cosmology and archaeology, researchers are tasked with navigating through the complexities of time and space to unearth ancient connections. Cosmic Archaeology demands a fusion of traditional archaeological methods with state-of-the-art cosmological tools, presenting a unique set of challenges and opportunities. One such approach involves the integration of astronomical data with archaeological findings, enabling the identification of celestial patterns encoded in ancient artifacts and structures. This methodology not only sheds light on ancient astronomical knowledge but also offers insights into the cultural significance of celestial events. Additionally, the utilization of advanced imaging technologies, such as ground-penetrating radar and LiDAR systems, allows for the non-invasive exploration of cosmic relics buried beneath terrestrial surfaces. By merging these disparate disciplines, researchers can reconstruct historical landscapes and uncover traces of extraterrestrial influence. Furthermore, isotopic analysis and radiometric dating techniques facilitate the temporal correlation between cosmic events and human activities, providing a chronological framework for understanding our intertwined cosmic heritage. Beyond terrestrial confines, remote sensing and spectroscopic analyses extend the reach of Cosmic Archaeology to celestial bodies, offering glimpses of potential extraterrestrial artifacts or environmental traces. The comparative study of material culture from diverse planetary bodies holds the promise of identifying common threads that link humanity with the cosmos. As Cosmic Archaeology continues to flourish, the development of interdisciplinary frameworks becomes imperative. Collaboration between experts in astronomy, anthropology, geology, and other relevant fields fosters the exchange of methodologies and best practices, fostering a holistic approach to cosmic exploration. Moreover, the integration of citizen science initiatives and public engagement programs empowers a wider community to contribute to the discovery and preservation of cosmic heritage. Ultimately, methodological innovation serves as the cornerstone of Cosmic Archaeology, propelling the quest to bridge cosmology and archaeology toward a deeper understanding of our place in the universe.

Case Studies Illustrating the Convergence

In the exploration of the intersection between cosmology and archaeology, case studies play a pivotal role in demonstrating the practical application of theoretical frameworks and methodological approaches. These case studies serve as real-world examples that showcase how the convergence of these disciplines can lead to profound insights into the history of our universe and the development of human civilizations. One such compelling case study involves the investigation of ancient celestial alignments in megalithic structures across various cultures. By combining astronomical observations with archaeological evidence, researchers have been able to discern intricate patterns that suggest an intimate connection between the architectural design of these structures and celestial phenomena, shedding light on the cosmological beliefs and practices of ancient societies. Another fascinating case study revolves around the analysis of cosmic dust particles found in archaeological sites, providing clues about extraterrestrial events that may have impacted Earth's ecosystems and human cultures. The examination of these microscopic cosmic remnants not only contributes to our understanding of cosmic processes but also offers valuable insights into the interplay between celestial events and human history. Furthermore, the study of ancient artifacts and petroglyphs depicting celestial bodies and astronomical events has unveiled cultural narratives that intertwine cosmological knowledge with societal beliefs and practices. This interdisciplinary approach has revealed how ancient civilizations interpreted and integrated celestial events into their cultural and religious frameworks, highlighting the entwined nature of cosmology and archaeology in shaping human perspectives. Additionally, the utilization of advanced imaging technologies, such as remote sensing and satellite mapping, has enabled the discovery and interpretation of ancient geoglyphs and earthworks that demonstrate an intricate knowledge of celestial alignments and astronomical phenomena. These spatial manifestations bear testament to the profound influence of cosmological understanding on the architectural and ceremonial landscape of ancient societies. Through a series of meticulously documented case studies, the convergence of cosmology and archaeology emerges as a transformative lens through which we can unravel the interconnected tapestry of cosmic evolution and human cultural development. These compelling investigations not only underscore the significance of interdisciplinary collaboration but also inspire further exploration into the rich nexus of cosmic history and human heritage.

Key Theories and Hypotheses

In the realm of cosmic archaeology, numerous key theories and hypotheses have emerged to guide research and exploration. One of the fundamental theories is the panspermia hypothesis, which postulates that life exists throughout the universe and can travel between celestial bodies. Panspermia proposes that the seeds of life, such as bacteria or simple organic molecules, may have been transported through space and seeded in new environments, including Earth. This hypothesis has significant implications for our understanding of the origins of life and the potential interconnectivity of life across the cosmos. Another pivotal theory is the Rare Earth hypothesis, which suggests that Earth-like planets capable of supporting complex terrestrial life are exceedingly rare in the universe. This theory takes into account the numerous factors that must align to create a habitable planet, including the right distance from a star, the presence of water, and a stable atmosphere. Understanding this hypothesis is crucial for evaluating the potential for extraterrestrial civilizations and the uniqueness of Earth as a cradle for life. Furthermore, the Fermi Paradox and the Drake Equation represent important theoretical frameworks in the search for extraterrestrial intelligence. The Fermi Paradox addresses the contradiction between the high probability of extraterrestrial civilizations existing and the lack of evidence for, or contact with such civilizations. Meanwhile, the Drake Equation serves as a hypothetical framework for estimating the number of active, communicative extraterrestrial civilizations in the Milky Way galaxy. These theories provide thought-provoking perspectives on the potential prevalence or scarcity of intelligent alien life within our cosmic neighborhood. Beyond these prominent theories, hypotheses related to technological signatures of advanced civilizations in space have gained increasing attention. The concept of Dyson spheres, megastructures that encapsulate a star to harness its energy, and other technosignatures serve as intriguing hypotheses in the quest to detect the presence of alien civilizations through astronomical observations. In parallel, the possibility of ancient extraterrestrial visitations shaping human history and cultural development has fueled hypotheses related to anomalous archaeological findings and potential traces of non-human influences. As researchers continue to delve into the depths of cosmic archaeology, these key theories and hypotheses form the foundation for inquiry, sparking new avenues of exploration and fueling interdisciplinary dialogues. They provoke contemplation on humanity's place in the universe, the potential diversity of life forms beyond Earth, and the intricate tapestry of cosmic phenomena that shape our understanding of existence itself.

Technological Advancements Facilitating New Discoveries

The exploration of the intersection between cosmology and archaeology has been significantly enhanced by a myriad of technological advancements, which have played an instrumental role in propelling discoveries and reshaping our understanding of the universe and ancient civilizations. One of the most pivotal technological breakthroughs lies in the realm of remote sensing techniques, such as LiDAR (Light Detection and Ranging), which has revolutionized how we uncover and map ancient landscapes and archaeological sites. Using airborne laser scanning technology, LiDAR has enabled researchers to penetrate dense forest canopies and visualize ancient features with unprecedented precision, revealing hidden structures previously inaccessible to traditional ground-based surveys. Furthermore, the application of advanced imaging technologies, including ground-penetrating radar (GPR) and multispectral satellite imagery, has facilitated the non-invasive examination and interpretation of underground archaeological remains, allowing for comprehensive assessments of historical landscapes without extensive excavation, thereby preserving these sites for future generations while still gleaning valuable insights. Integrating these innovative tools has proven invaluable in illuminating long-lost civilizations and uncovering their interactions with cosmic phenomena. In addition to remote sensing, developments in analytical instrumentation have substantially heightened the capabilities of archaeological investigations. Mass spectrometry, for instance, has enabled precise isotopic analysis of unearthed materials, offering crucial evidence concerning resource procurement, trade routes, and human migration patterns throughout history. Meanwhile, high-resolution microscopy and elemental mapping techniques have unveiled intricate details within artifacts, elucidating craftsmanship, provenance, and material composition, creating a comprehensive picture of ancient societies' ingenuity and cultural exchanges. Moreover, the burgeoning field of computational modeling and simulation has empowered researchers to reconstruct and simulate ancient environments, celestial events, and astronomical alignments, providing deeper contextualization of archaeological findings within the broader cosmic narrative. With the aid of advanced 3D visualization software and virtual reality platforms, scholars and enthusiasts alike can immerse themselves in meticulously reconstructed ancient landscapes and celestial configurations, fostering a heightened appreciation for the interconnectedness of human heritage and the cosmos. As we continue to witness remarkable technological progress, prospects for uncovering and comprehending the cosmic tapestry of our past burgeon, underscoring the transformative influence of these innovations on the collaborative pursuit of knowledge at the nexus of cosmology and archaeology.

Challenges and Controversies in Interdisciplinary Studies

Interdisciplinary studies at the intersection of cosmology and archaeology present a myriad of challenges and controversies that require careful consideration and strategic navigation. One of the primary challenges lies in the inherent complexity of merging distinct scientific methodologies, terminologies, and epistemological frameworks. The disparities like evidence, from astronomical observations to material culture, often lead to contrasting interpretation paradigms, raising questions about the compatibility of diverse data sources. Addressing this challenge involves fostering a deep understanding of the underlying principles in both fields while developing new interdisciplinary methodologies for analysis and interpretation. Another fundamental issue pertains to the allocation of resources and funding for interdisciplinary research endeavors. Traditional funding models and academic structures may not readily accommodate the collaborative nature of cosmic archaeology, necessitating innovative approaches to garner support and recognition for this amalgamation of disciplines. Additionally, as interdisciplinary studies transcend the boundaries of individual disciplines, they face resistance from purists who may view such integration as diluting the rigor and authenticity of their respective fields. Overcoming these contentions requires effective communication and advocacy to demonstrate the unique insights and transformative potential that arise from the synergy of cosmology and archaeology. Moreover, ethical considerations stand as a significant point of contention in interdisciplinary studies, particularly when engaging with materials and narratives related to ancient civilizations and potential extraterrestrial influences. Balancing the preservation of cultural heritage with the pursuit of scientific inquiry demands a delicate ethical framework that acknowledges and respects diverse cultural perspectives, heritage rights, and the potential impacts of discoveries. This necessitates collaboration with indigenous communities, ethical review boards, and cultural heritage authorities to ensure responsible and respectful research practices. Furthermore, the interdisciplinary nature of cosmic archaeology raises methodological controversies, especially concerning the integration of qualitative and quantitative approaches. Negotiating the differing standards for evidence assessment, inference, and validation requires the establishment of hybrid methodologies that leverage the strengths of both disciplines while mitigating their respective limitations. Transparency and reflexivity in methodological decision-making serve as crucial elements in navigating these controversies and advancing the credibility and robustness of interdisciplinary findings. In conclusion, the challenges and controversies inherent in interdisciplinary cosmic archaeology underscore the necessity of fostering open dialogue, principled negotiation, and continuous adaptation. By acknowledging and addressing these complexities, the field can chart a path toward transformative discoveries that resonate across multiple domains of human knowledge and experience.

Future Directions in Hybrid Research

As we stand at the precipice of groundbreaking discoveries, the future of hybrid research in cosmology and archaeology beckons with boundless potential. The synergistic collaboration between these two disciplines offers a path toward unraveling the enigmatic tapestry of our cosmic origins and civilizations. Looking ahead, one of the foremost objectives in hybrid research is to leverage advanced computational modeling and data analysis techniques to glean deeper insights from the vast troves of cosmic and archaeological data. Integrating artificial intelligence and machine learning algorithms holds promise in identifying patterns and correlations that may have eluded previous methodologies, paving the way for transformative revelations. In the realm of space exploration, the burgeoning field of astroarchaeology is poised to revolutionize our understanding of ancient extraterrestrial encounters and potential artifacts residing beyond Earth. By integrating space missions and archaeological methodologies, scientists are on the brink of uncovering relics or remnants that could provide tantalizing evidence of interactions with otherworldly entities. Furthermore, the development of cutting-edge instrumentation, such as advanced spectroscopy and imaging technologies, is set to enhance our ability to detect and analyze celestial phenomena with unparalleled precision, thereby opening new frontiers in uncovering cosmic archeological vestiges. The advent of interdisciplinary research centers focused on cosmic archaeology will serve as crucibles for fostering collaboration and propelling the field towards innovative horizons. These hubs will incubate interdisciplinary teams comprising astrophysicists, archaeologists, data scientists, and technologists, engendering a melting pot of ideas and expertise. Additionally, these institutions will play a pivotal role in nurturing the next generation of researchers who possess a holistic perspective and skill set tailored for the demanding landscape of cosmic exploration. A paramount challenge on this trajectory is mitigating knowledge silos and fostering effective communication across different scientific domains. Establishing common ground and shared vocabularies will be imperative for transcending disciplinary barriers and harnessing the collective intellect and resources of diverse research communities. Interdisciplinary workshops, symposiums, and collaborative projects will fortify the interconnectedness of cosmology and archaeology, ultimately catalyzing a harmonious integration of techniques and concepts for comprehensive cosmic inquiries. The horizon of hybrid research is also marked by a resolute commitment to ethical considerations in examining the implications of potential discoveries. As we journey towards unearthing profound revelations about our place in the cosmos, it becomes incumbent upon us to contemplate the sociocultural, philosophical, and existential ramifications that may ensue. Embracing an ethical framework that upholds respect for ancient cultures, Indigenous knowledge systems, and potential extraterrestrial life forms will guide the trajectory of hybrid research while fostering global cooperation and inclusivity. In conclusion, the emerging frontiers of hybrid research in cosmology and archaeology are underpinned by an unwavering dedication to collaborative ingenuity, technological innovation, and ethical stewardship. The synthesis of these disciplines heralds an era of unprecedented discovery, where the mysteries of our cosmic heritage may be unveiled, transcending temporal and spatial boundaries.

Conclusion: The Significance of Bridging Disciplines

As we conclude our exploration of bridging disciplines in cosmic archaeology, it is imperative to underscore the profound significance of integrating cosmology and archaeology towards advancing our understanding of the universe's enigmatic past. The convergence of these disciplines has not only broadened our intellectual horizons but has also expanded the scope of scientific inquiry into uncharted territories. By reconciling the macroscopic narratives of cosmology with the microscopic insights of archaeology, we have embarked on an exhilarating journey that transcends conventional academic boundaries. This interdisciplinary synergy has fostered a holistic approach that encapsulates the interconnectedness of celestial phenomena and the terrestrial remnants of ancient civilizations. Through this amalgamation, we have witnessed the emergence of revolutionary perspectives that challenge preconceived notions, unveiling new dimensions of inquiry and discovery. Moreover, the collaborative efforts between cosmologists and archaeologists have yielded a symbiotic exchange of methodologies, propelling the development of novel tools and techniques for investigating cosmic artifacts and extraterrestrial relics. The significance of bridging disciplines lies not only in the illumination of historical intricacies but also in the contemplation of our place within the cosmic tapestry. It prompts us to ponder profound questions about our origins, purpose, and the potential existence of other sentient beings beyond the confines of our planet. Such contemplation ignites a sense of shared curiosity and wonder that transcends cultural and geographical divides, compelling humanity to delve deeper into the enigmas of the cosmos. Moreover, the synthesis of cosmology and archaeology engenders a paradigm shift in our philosophical and ethical frameworks. It invokes a reevaluation of our stewardship over the vast cosmic landscape and underscores the ethical responsibilities entailed in our pursuit of cosmic knowledge. As we navigate this uncharted terrain, it becomes increasingly evident that our quest for understanding necessitates a harmonious coalescence of scientific rigor, ethical considerations, and global collaboration. For the aspiring cosmic archaeologist, this convergence beckons the call to embark on a daring odyssey— one that interweaves empirical inquiry, speculative imagination, and unwavering determination. Embracing the significance of bridging disciplines heralds an era of unprecedented intellectual richness and transdisciplinary synergy, propelling us toward a future where the mysteries of the cosmos are unraveled through the collective wisdom of diverse fields of inquiry. As we stand at the crescendo of this remarkable narrative, let us remain steadfast in our commitment to transcend disciplinary confines and venture forth into the profound, interconnected realms of cosmology and archaeology.

Chapter Two

The Cosmic Enigma

Unveiling the Enigma: An Overview

Cosmic enigmas, spanning the vast expanse of the universe, have perplexed and inspired scientists, philosophers, and thinkers throughout history. Defined as phenomena that defy current scientific understanding and explanation, cosmic enigmas encompass a broad spectrum of mysteries, from the nature of dark matter and energy to the origins of celestial bodies and cosmic phenomena. These enigmas transcend disciplinary boundaries, integrating insights from astronomy, physics, cosmology, and even archaeology. Within the historical context, ancient civilizations and cultures often grappled with their understanding of the cosmos, leading to the development of mythologies, creation narratives, and early attempts at cosmic interpretation. As scientific knowledge evolved, these enigmatic aspects of the universe became focal points of inquiry for modern researchers, prompting innovative theories, hypotheses, and analytical frameworks. The scope of cosmic enigmas extends from the macroscopic realm of galaxies and cosmic structures to the microscopic domain of quantum mechanics and subatomic particles. By scrutinizing these phenomena, scholars aim to decipher the fundamental laws governing the cosmos and unravel the mysteries that have eluded human comprehension for millennia. Across different scientific disciplines, the exploration of cosmic enigmas unveils complex interconnections, as researchers draw upon diverse methodologies, technologies, and theoretical paradigms to investigate these pervasive mysteries. Thus, the quest to unravel the cosmic enigma gives rise to an interdisciplinary endeavor that transcends traditional boundaries, inviting collaboration and exchange among experts in various fields. This cross-disciplinary approach enriches the analysis of cosmic enigmas, fostering a comprehensive and multifaceted understanding of the universe's elusive phenomena. Ultimately, the comprehensive exploration of the historical context of cosmic mysteries amplifies the significance of these enigmas, shedding light on humanity's perpetual pursuit of knowledge about the cosmos and unveiling the intricate tapestry of questions that continue to challenge contemporary scientific thought.

The Historical Context of Cosmic Mysteries

Throughout the annals of human history, civilizations across the globe have gazed upon the cosmos with wonder and curiosity, seeking to comprehend its enigmatic nature. From the ancient civilizations of Mesopotamia and Egypt to the Renaissance thinkers and modern astronomers, the quest to unravel the mysteries of the universe has been a driving force behind human intellectual endeavor. The historical context of cosmic mysteries encompasses a rich tapestry of mythologies, philosophical musings, and scientific inquiry that have shaped our understanding of the cosmos. Early civilizations, such as the Sumerians and Babylonians, observed the celestial movements and endeavored to interpret them through intricate cosmological beliefs, linking the heavens to the earth and human destiny. These early interpretations influenced the development of astronomy, laying the foundation for future astronomical endeavors. During the Renaissance, visionaries like Copernicus and Galileo challenged prevailing geocentric theories, igniting a scientific revolution that paved the way for modern cosmology. The historical lineage of cosmic mysteries extends into the contemporary era, marked by groundbreaking discoveries and paradigm-shifting theories that continue to redefine our relationship with the universe. The evolution of cosmic mysteries is inherently intertwined with the evolution of human thought, and unravelling this historical narrative provides valuable insights into the deep-rooted human fascination with the cosmic enigma. By examining the historical context of cosmic mysteries, we can gain a profound appreciation for the diverse perspectives and enduring quests that have propelled humanity's engagement with the cosmos, culminating in the interdisciplinary field of cosmic archaeology.

Fundamental Cosmos Theories and Their Implications for Archaeology

The intersection of fundamental cosmos theories and archaeology presents an intriguing landscape of scientific inquiry and exploration. One cannot delve into the mysteries of cosmic origins without considering the profound implications of these theories on the field of archaeology. At its core, the study of fundamental cosmos theories encompasses a wide range of concepts and principles, including the Big Bang theory, cosmological inflation, dark matter, and the cosmic microwave background radiation. Each of these theories provides a unique lens through which we can interpret the origins, evolution, and structure of the universe. As these fundamental concepts continue to evolve and expand, their implications for archaeology become increasingly pronounced. The interplay between cosmic theories and archaeological investigations holds immense potential for unraveling the enigmatic tapestry of our cosmic heritage. From redefining our understanding of ancient civilizations to shedding light on the emergence of life beyond Earth, fundamental cosmos theories offer a wealth of insights that can profoundly shape the trajectory of archaeological research. The Big Bang theory, for instance, not only elucidates the origin of the universe but also informs our understanding of the cosmic materials that eventually coalesced to form celestial bodies, including planets hospitable to life. This temporal framework has significant ramifications for the study of ancient artifacts and geological formations on Earth, providing crucial context for interpreting the deep-time history of our planet. The concept of dark matter, another cornerstone of modern cosmology, introduces a compelling dimension to our comprehension of cosmic structures and gravitational interactions. In the realm of archaeology, this enigmatic cosmic ingredient prompts us to consider the invisible forces that may have shaped the distribution and movement of celestial resources, impacting the development and dispersion of ancient human societies. Moreover, the cosmic microwave background radiation, often regarded as the echo of the primordial universe, offers invaluable clues about the early composition and dynamics of the cosmos. Its relevance to archaeology lies in the potential for linking cosmic events to terrestrial phenomena, providing a nuanced perspective on the entwined histories of the universe and our planet. By delving into these foundational cosmos theories, archaeologists gain a deeper appreciation for the interconnectedness of cosmic processes and terrestrial phenomena, fostering a holistic approach to unraveling the mysteries of ancient civilizations and extraterrestrial influences. As technology advances and our understanding of fundamental cosmos theories continues to deepen, the opportunities for collaborative research between cosmologists and archaeologists are boundless, fostering a symbiotic relationship that promises to enrich both fields with novel insights and discoveries.

Analytical Methods in Cosmic Pattern Analysis

Cosmic pattern analysis entails a rigorous application of analytical methods to uncover intricate connections and recurring motifs within the cosmic landscape. At the heart of this endeavor lies the utilization of advanced computational algorithms, statistical models, and data visualization techniques to discern patterns from an ocean of cosmic information. Through the lens of archaeology, these analytical methods serve as powerful tools for unraveling enigmatic traces left by ancient cosmic phenomena.

One such method is spectral analysis, which allows us to decompose complex cosmic signals into their constituent frequencies. By examining the spectral signatures of celestial events, we can glean insights into their underlying mechanisms and potential implications for terrestrial archaeological findings. Coupled with sophisticated imaging and spectroscopic technologies, this approach provides a window into the distant past, enabling researchers to identify patterns that transcend both time and space.

Furthermore, fractal analysis plays a pivotal role in characterizing cosmic structures exhibiting self-similar features across different scales. This approach offers a means to quantify the complexity and irregularity present in cosmic artifacts and phenomena, shedding light on underlying dynamics and potential relationships with terrestrial archaeological contexts. As cosmic archaeologists venture into realms where classical geometric forms often fail to apply, fractal analysis emerges as an indispensable tool for comprehending the bewildering intricacies of the cosmic enigma.

In addition, Bayesian inference, a cornerstone of modern statistical analysis, empowers cosmic archaeologists to assess the likelihood of various cosmic patterns and their associated interpretations. This probabilistic framework enables the integration of prior knowledge, observational data, and theoretical models, providing a robust foundation for evaluating the significance of cosmic alignments, geometric arrangements, and anomalous configurations. By harnessing the power of Bayesian reasoning, researchers can navigate through the labyrinth of cosmic patterns with enhanced clarity and precision.

Ultimately, the fusion of these analytical methods catalyzes a transformative synergy between physics and archaeology, illuminating cryptic threads of cosmic continuity that bind our origins to the farthest reaches of the cosmos. It is through the relentless pursuit of understanding cosmic patterns that we unveil new avenues for exploring the enigmatic tapestry of our cosmic heritage, transcending traditional disciplinary boundaries to unlock the timeless mysteries that have shaped our existence.

Interdisciplinary Bridges: Combining Physics with Archaeology

Interdisciplinary collaboration plays a crucial role in the field of cosmic archaeology, particularly in the intersection of physics and archaeology. The integration of these disciplines provides a unique opportunity to unravel the mysteries of our cosmic origins, ancient civilizations, and extraterrestrial phenomena. By merging the technical expertise of physicists with the contextual understanding of archaeologists, we can gain unprecedented insights into the cosmic enigma.

At the core of this interdisciplinary endeavor lies the utilization of advanced scientific tools and methodologies from physics to analyze archaeological artifacts. Techniques such as spectroscopy, radiocarbon dating, and complex imaging modalities allow us to probe the material composition, age, and cultural significance of unearthed relics in a manner that transcends traditional archaeological approaches. By applying principles of quantum mechanics and astrophysics to ancient materials, scientists can discern subtle traces of cosmic influences embedded within archaeological remnants, shedding light on our celestial connections.

Furthermore, the synergy between physics and archaeology extends beyond the realm of material analysis, encompassing the study of cosmic events and their impact on human history. Integrating astrophysical models with archaeological data enables us to pinpoint pivotal cosmic occurrences, such as meteorite impacts or astronomical phenomena, that have left indelible imprints on ancient societies. By understanding the cosmic context in which ancient civilizations flourished, we can reconstruct narratives of cosmic encounters and elucidate the sociocultural implications of celestial events.

Moreover, the collaboration between physicists and archaeologists fosters the development of innovative technologies tailored for cosmic archaeology. From advanced remote sensing techniques for detecting extraterrestrial remains to sophisticated cosmological simulations that contextualize archaeological findings within cosmic timelines, the fusion of physics and archaeology propels the field towards unparalleled methodological advancements. This convergence of expertise not only expands the scope of cosmic archaeology but also enriches both disciplines, encouraging novel approaches and perspectives in the pursuit of unraveling cosmic mysteries.

In essence, the interdisciplinary bridges established by combining physics with archaeology propel the field of cosmic archaeology into uncharted frontiers of knowledge. Through collaborative endeavors, researchers confront the cosmic enigma with multifaceted insights, leveraging the collective wisdom of diverse disciplines to decode the intricate tapestry of our cosmic heritage.

Key Cosmic Events and Their Archaeological Significance

The history of cosmic events is a narrative filled with epochs of significant happenings that have left an indelible mark on the universe. From the primordial nucleosynthesis during the early moments after the Big

Bang to the celestial cataclysms that shaped the formation of galaxies, each cosmic event has contributed to the composition and development of our cosmos. These events not only hold intrinsic value for astrophysicists and cosmologists but also bear profound relevance to the field of archaeology. By delving into the archaeological record, researchers can discern tangible remnants or effects of these cosmic occurrences on Earth.

One of the key cosmic events with a palpable impact on terrestrial history is the collision of celestial bodies with our planet. Meteorite impacts, in particular, have had far-reaching consequences that are imprinted in the geological and biological archives. The study of impact craters around the world provides vital clues about the environmental upheavals caused by such cosmic collisions. Archaeological evidence corroborates the link between catastrophic events and abrupt changes in ancient societies, offering insights into how these civilizations coped with and adapted to the aftermath of such cataclysms.

Furthermore, the phenomenon of supernovae—the explosive deaths of massive stars—has not only influenced the chemical composition of the universe but has also left a discernible mark on Earth. Traces of supernova remnants have been found within terrestrial materials, shedding light on the pervasive effects of stellar explosions on our planet. Through meticulous analyses of archaeological artifacts, scientists have uncovered intriguing correlations between radiocarbon variations and supernova events, hinting at the potential influence of stellar phenomena on human cultures and historical timelines.

Another pivotal cosmic event of archaeological significance is the cyclical variations in Earth's climate, often triggered by astronomical factors such as Milankovitch cycles. These climatic fluctuations have played an instrumental role in shaping human migration patterns, agricultural practices, and societal developments throughout history. By examining the relics and settlements of past civilizations, archaeologists can discern the adaptive strategies employed by ancient populations in response to changing climatic conditions, providing valuable lessons for contemporary global challenges.

In essence, the intersections between cosmic events and archaeology offer a captivating avenue for exploring the interconnectedness of the universe and its terrestrial ramifications. By unraveling the archaeological significance of key cosmic events, researchers not only enrich our understanding of the past but also illuminate the enduring influence of cosmic phenomena on human history and cultural evolution.

Controversies and Debates in Modern Cosmology

Modern cosmology is a field rife with controversies and debates that have captured the attention of scientists, scholars, and the public alike. At the heart of these discussions are fundamental questions about the nature of the universe, its origins, and its ultimate fate. One of the most hotly debated topics is the issue of dark matter and dark energy, which together are thought to make up over 95% of the total mass-energy content of the universe. Despite extensive research efforts, the true nature of these enigmatic components remains elusive, leading to a palpable sense of uncertainty within the scientific community. Moreover, competing theories such as the inflationary model and alternative gravitational theories have sparked intense debates regarding the validity of existing paradigms. These discussions often confront researchers with deeply philosophical conundrums, raising profound questions about the nature of scientific knowledge and the boundaries of human understanding. The controversies in modern cosmology also extend to the realm of cosmic archaeology, where interpretations of astronomical data and cosmic events can vary widely, leading to ongoing debates about the implications for our understanding of ancient civilizations and extraterrestrial influences. The potential discovery of extraterrestrial life and the search for technological signatures of advanced alien civilizations further contribute to the contentious landscape of modern cosmology. Ethical considerations, societal implications, and the impact on existing worldviews add layers of complexity to these debates, intensifying the intellectual and emotional stakes involved. As researchers continue to grapple with these controversies, it becomes increasingly clear that cosmology is not only an empirical science but also a crucible for philosophical, ethical, and existential inquiries. Navigating these treacherous waters demands a delicate balance between rigorous scientific inquiry and open-minded contemplation, ensuring that the pursuit of cosmic truths remains an intellectually vibrant and morally responsible endeavor.

Case Studies: Unexplained Phenomena and Artifacts

In this section, we delve into compelling case studies that showcase unexplained phenomena and artifacts with potential cosmic significance. These case studies offer invaluable insights into the enigmatic intersection of archaeology and cosmology, compelling readers to contemplate the profound implications of unearthed mysteries. One such case study revolves around an ancient artifact discovered in a remote archaeological site, exhibiting anomalous properties that defy conventional scientific explanations. Through meticulous analysis and interdisciplinary collaboration, researchers have traced the artifact's origins to a cosmic event predating human civilization, prompting profound questions about extraterrestrial influences on ancient cultures. Another riveting case study explores unexplained phenomena recorded in ancient texts and oral traditions from disparate civilizations across the globe. These accounts allude to inexplicable celestial events and encounters with otherworldly beings, challenging prevailing paradigms of human history and igniting fervent debates within the scientific community. By scrutinizing these enigmatic accounts through a cosmological lens, researchers aim to unravel the underlying truths embedded in ancient narratives and folklore. Furthermore, contemporary satellite imagery has unveiled perplexing geometric formations etched into remote landscapes, evoking tantalizing parallels with celestial constellations and spatial configurations. The convergence of these enigmatic terrestrial patterns with cosmic symmetries raises thought-provoking implications about potential extraterrestrial interventions in Earth's ancient past. Through meticulous fieldwork and advanced imaging technologies, scientists seek to discern the enigmatic origins and purpose behind these enigmatic terrestrial features, unlocking the cryptic interplay between cosmic forces and terrestrial manifestations. Ultimately, these captivating case studies invite readers to embark on a journey of intellectual exploration, inviting them to ponder the intricacies of past cosmic encounters and their enduring impact on humanity's collective consciousness.

The Impact of Recent Discover Emerges Analyses in Classical Theories

As our understanding of the cosmos deepens, the impact of recent discoveries on classical theories becomes increasingly significant. With advancements in technology and observations, scientists and archaeologists have gained new insights into the nature of cosmic phenomena and its implications for ancient civilizations. One such impact is the reevaluation of existing theoretical frameworks in light of groundbreaking findings. Examination of cosmic relics and artifacts has led to a reassessment of established archaeological and cosmological paradigms, challenging traditional notions of human history and the origins of the universe. By integrating data from diverse fields such as astronomy, astrophysics, chemistry, and archaeology, researchers have been able to develop more comprehensive models that reflect the intricate relationship between cosmic evolution and terrestrial cultures. Moreover, the analysis of cosmic patterns and anomalies has unearthed compelling evidence that prompts a revision of long-standing assumptions regarding the formation of celestial bodies and their role in shaping early human societies. These recent discoveries not only enrich our knowledge of the cosmos but also foster a deeper appreciation for the enduring interplay between cosmic forces and human endeavors.

Furthermore, the impact of these analyses extends beyond the realm of scientific inquiry, influencing philosophical discourses and inspiring a re-examination of fundamental existential questions. The profound implications of recent discoveries in cosmic archaeology serve as a catalyst for rethinking humanity's position within the cosmic tapestry, prompting contemplation of our place in the vast expanse of the universe. This transformative process invites scholars and enthusiasts alike to ponder the interconnectedness of cosmic phenomena and human experiences, encouraging a holistic approach to understanding the mysteries of existence. Collaborative efforts among multidisciplinary teams lay the groundwork for innovative investigations that transcend conventional disciplinary boundaries, illuminating novel pathways for unlocking the secrets of the cosmos.

In conclusion, the impact of recent discoveries and analyses in classical theories marks a pivotal phase in the evolution of cosmic archaeology. By revisiting established concepts and integrating new perspectives, the field continues to evolve, uncovering hidden facets of the cosmic enigma and reshaping our perception of the universe. As we embark on this journey of exploration and discovery, the convergence of archaeological and cosmological inquiry paves the way for a harmonious fusion of ancient wisdom and modern insight, propelling us toward a deeper comprehension of the cosmic narrative.

Looking Forward: Emerging Trends and Predictions in Cosmic Archaeology

As the field of cosmic archaeology continues to evolve, there are several emerging trends and exciting predictions that hold promise for unraveling the mysteries of our cosmic origins. One key trend is the increasing integration of advanced technology into archaeological research, allowing for more precise and nuanced analysis of extraterrestrial artifacts and phenomena. The development of sophisticated imaging techniques and remote sensing technologies promises to revolutionize our ability to study cosmic relics with unprecedented detail, providing invaluable insights into ancient civilizations beyond Earth.

Furthermore, the growing collaboration between cosmologists, astrophysicists, and archaeologists is fostering interdisciplinary approaches that transcend traditional boundaries. By leveraging the expertise of multiple scientific disciplines, researchers are poised to make revolutionary breakthroughs in understanding the connections between cosmic events and their potential influence on terrestrial cultures throughout history. This collaborative synergy is likely to propel cosmic archaeology into an era of unprecedented discovery and innovation.

Advancements in astrobiological research also hold tremendous potential for cosmic archaeology. With ongoing efforts to identify habitable exoplanets and search for signs of extraterrestrial life, the prospect of discovering ancient traces of alien civilizations becomes increasingly plausible. The intersection of astrobiology and cosmic archaeology is poised to open new frontiers in our quest to comprehend the diversity of life in the cosmos and its impact on the development of intelligent beings elsewhere.

Looking ahead, the burgeoning field of space archeology presents an intriguing frontier for exploration. By extending archaeological principles to space exploration, researchers may uncover evidence of ancient extraterrestrial technologies or remnants of interstellar civilizations, fundamentally reshaping our understanding of the universe's history. Furthermore, the proliferation of private space companies and ambitious missions to celestial bodies offer unprecedented opportunities for cosmic archaeological discoveries beyond the confines of Earth.

These emerging trends collectively herald a future where cosmic archaeology stands at the forefront of scientific inquiry, poised to unravel profound truths about the nature of our existence in the cosmos. As we venture into this uncharted territory of discovery, anticipation mounts for the transformative revelations and groundbreaking insights that await us, promising to reshape our conceptualization of the cosmic enigma.

Chapter Three

The Foundations of Cosmic Archaeology

Defining Cosmic Archaeology

Cosmic archaeology, a burgeoning interdisciplinary field at the nexus of astronomy, anthropology, and archaeology, encapsulates the systematic study of extraterrestrial artifacts, cosmic relics, and the material remnants of ancient civilizations beyond Earth. Unlike traditional archaeological endeavors confined to terrestrial landscapes, cosmic archaeology expands the scope of inquiry to encompass celestial bodies, interstellar phenomena, and potential traces of extraterrestrial intelligence. This multidimensional exploration seeks to unravel the enigmatic footprints of ancient cosmic inhabitants, decipher their technological vestiges, and fathom the profound implications for our comprehension of the universe. At the heart of cosmic archaeology lies the pivotal quest to uncover artifacts, structures, or remnants that may offer glimpses into the existence of advanced extraterrestrial societies, astrophysical events, or interactions between terrestrial and cosmic entities. As humanity increasingly turns its gaze towards the cosmos in pursuit of answering existential questions, cosmic archaeology emerges as an indispensable tool for augmenting our understanding of the intricate tapestry of the universe. Its ramifications extend far beyond the mere retrieval of tangible evidence, transcending into the realms of philosophy, ethics, and the very essence of sentient existence. By defining cosmic archaeology, we embark on a profound journey that redefines conventional notions of archaeological inquiry, expanding the horizons of exploration and empowering us to contemplate the unfathomable possibilities intertwined with cosmic heritage.

Historical Overview of Cosmic Archaeology

The emergence of cosmic archaeology as a scientific discipline is deeply rooted in humanity's innate curiosity about the cosmos and our place within it. While traditional archaeology has long focused on uncovering and interpreting human history through material remains on Earth, the concept of cosmic archaeology represents a paradigm shift that expands the scope of archaeological inquiry beyond the confines of our planet. The historical trajectory of cosmic archaeology can be traced back to ancient civilizations that exhibited an advanced understanding of celestial phenomena and potential extraterrestrial influences. From the architectural precision of ancient structures aligned with astronomical events to the preservation of cosmic symbolism in art and artifacts, early evidence suggests that humans have contemplated cosmic connections for millennia. However, the formalization of cosmic archaeology as a distinct field gained momentum during the 20th century, driven by advances in space exploration, astronomy, and interdisciplinary collaboration. Pioneering scientists and thinkers began questioning whether our understanding of past civilizations and their interactions with the cosmos should be confined to terrestrial boundaries. This sparked a reevaluation of archaeological methodologies and theories to encompass the broader context of our cosmic environment. The advent of satellite technology, telescopic observations, and space missions provided unprecedented access to extraterrestrial landscapes and celestial bodies, further igniting the pursuit of cosmic archaeological inquiry. Notably, the realization that cosmic debris, such as meteorites and comets, could bear insights into both Earth's ancient past and the wider cosmic narrative sparked new interest in integrating cosmic and terrestrial archaeological practices. The historical overview of cosmic archaeology underscores the evolving interplay between scientific curiosity, technological progress, and our enduring fascination with the cosmos. As we delve deeper into the historical foundations of cosmic archaeology, we gain a profound appreciation for the interconnectedness of human history, celestial dynamics, and the vast, enigmatic tapestry of the universe.

Methodologies and Tools in Cosmic Archaeological Research

Cosmic archaeological research employs a wide array of sophisticated methodologies and cutting-edge tools to unravel the mysteries of extraterrestrial influence on our planet and beyond. An interdisciplinary approach is crucial, integrating techniques from archaeology, astronomy, astrophysics, and even astrobiology to comprehensively study ancient artifacts, sites, and cosmic phenomena.

Remote sensing techniques play a pivotal role in cosmic archaeological research, enabling scientists to survey vast areas and identify potential sites of interest. Advanced satellite imagery, LiDAR technology, and ground-penetrating radar facilitate the identification and mapping of ancient structures and geological features that may hold clues to extraterrestrial interactions in the distant past.

Furthermore, state-of-the-art laboratory analyses are integral to cosmic archaeological investigations. Radiocarbon dating, thermoluminescence dating, and other dating methods help establish the chronology of unearthed artifacts, providing invaluable insights into their origins and potential connections to cosmic events. Additionally, elemental and isotopic analyses offer clues about the composition and provenance of materials, shedding light on their possible extraterrestrial associations.

The use of high-resolution imaging technologies, such as 3D scanning and microscopy, allows for detailed examinations of intricate artifact designs and inscriptions, enabling researchers to discern patterns and symbols that may reflect celestial influences or advanced technological knowledge from extraterrestrial sources.

In parallel, computational modeling and simulation tools have revolutionized cosmic archaeological research, facilitating the reconstruction of ancient planetary alignments, astronomical observations, and potential celestial events that could have influenced ancient civilizations. These tools also aid in deciphering enigmatic artifacts with possible astronomical significance.

Moreover, the integration of spectroscopic analyses enables scientists to uncover traces of extraterrestrial materials, isotopic anomalies, and chemical signatures that deviate from terrestrial norms, thus supporting hypotheses of cosmic interventions in human history.

Emerging technologies, such as machine learning algorithms and artificial intelligence, are increasingly applied to sift through vast datasets, identify patterns, and discern correlations between archaeological findings and celestial phenomena, providing new perspectives on the intersection of cosmic and terrestrial realms.

Ultimately, the fusion of these diverse methodologies and tools underpins the evolving landscape of cosmic archaeological research, propelling the quest to unveil the impact of extraterrestrial forces on our ancient past and contemplating their implications for humanity's future.

The Role of Multidisciplinary Approaches

As our understanding of cosmic archaeology continues to evolve, it becomes increasingly evident that investigating the mysteries of our universe demands a multidisciplinary approach. This complex and interconnected field necessitates the collaboration of experts from various scientific domains, including astronomy, astrophysics, geology, chemistry, biology, archaeology, and anthropology, among others. Each discipline contributes unique perspectives, methodologies, and tools that are essential for unraveling the intricate tapestry of cosmic artifacts and extraterrestrial phenomena. Through the convergence of diverse knowledge bases and research techniques, multidisciplinary approaches in cosmic archaeology help to contextualize findings, validate hypotheses, and foster innovative discoveries. An interdisciplinary framework enhances the robustness and reliability of research outcomes by ensuring that insights are corroborated through multiple lenses. By harnessing the collective expertise of different fields, researchers can more effectively synthesize and interpret data while mitigating biases inherent to singular disciplines. Moreover, the synergy of multidisciplinary collaboration facilitates holistic interpretations, allowing for a comprehensive understanding of cosmic phenomena that transcends compartmentalized silos of knowledge. The fusion of disparate academic realms also cultivates novel methodologies and instruments specifically tailored for cosmic archaeological endeavors, propelling the field towards greater efficacy and precision. Interdisciplinary collaboration broadens the scope of inquiry, enabling researchers to explore cosmic artifacts and enigmatic traces of extraterrestrial influence with depth and nuance. By drawing upon diverse disciplinary approaches, cosmic archaeologists can tap into a far-reaching reservoir of expertise and analytical tools, thereby enriching the depth and breadth of their investigations. Furthermore, the interplay between multidisciplinary teams encourages intellectual cross-pollination, sparking innovative ideas and fresh perspectives within the field.

This collaborative ethos catalyzes the evolution of cosmic archaeology, elevating the standard of research and nurturing a culture of continuous exploration and discovery. Ultimately, the integration of multidisciplinary approaches fortifies the foundation of cosmic archaeology, empowering investigators to confront the complexities of the universe with enhanced rigor, clarity, and ingenuity.

Key Concepts and Theories in Cosmic Archaeology

Cosmic archaeology is founded upon a rich tapestry of key concepts and theories that provide the framework for understanding the extraterrestrial influences on our past and present. At its core, cosmic archaeology delves into the intersection of space exploration, archaeology, and anthropology, unraveling the intricate network of interactions between terrestrial civilizations and potential extraterrestrial entities. This chapter will explore the fundamental concepts and theories that underpin cosmic archaeology, shedding light on the complex web of ideas that drive this emerging field of research.

One pivotal concept in cosmic archaeology is the notion of panspermia, which proposes that life may have originated from outer space and been transported to Earth, or vice versa. This theory challenges traditional views of terrestrial evolution and has profound implications for our understanding of the cosmic origins of life. Additionally, the concept of interstellar communication, including the search for extraterrestrial intelligence (SETI), is integral to cosmic archaeology, as it shapes our quest to decipher potential interactions with alien civilizations throughout history and across the cosmos.

Moreover, the Fermi paradox forms a cornerstone in the realm of cosmic archaeology, raising thought-provoking questions about the apparent absence of widespread extraterrestrial civilizations despite the high probability of their existence. This paradox fuels intense debate and speculation, driving researchers to explore diverse hypotheses and scenarios that may elucidate this enigmatic conundrum.

In parallel, the anthropic principle and the concept of cosmic fine-tuning are essential considerations in cosmic archaeology, as they contemplate the remarkable coalescence of physical constants and conditions necessary for the emergence of life in the universe. These principles prompt contemplation of the inherent connections between cosmic phenomena and the development of intelligent life forms, giving rise to profound philosophical and scientific inquiries.

Furthermore, the study of ancient astronaut theory, which proposes that intelligent extraterrestrial beings have interacted with human civilization since ancient times, offers an intriguing lens through which to analyze archaeological findings and historical narratives. Evaluating this theory within the context of cosmic archaeology demands a balanced exploration of evidence and speculation, stimulating critical discourse within the academic community.

Lastly, the concept of astrobiology is closely intertwined with cosmic archaeology, as it engenders interdisciplinary investigations into the potential diversity of life beyond Earth, probing the boundaries of our understanding of life's possible manifestations in the cosmos. By embracing these key concepts and theories, cosmic archaeology navigates the converging currents of scientific inquiry, aesthetic contemplation, and speculative exploration, opening new vistas of intellectual discovery and redefining the frontiers of human knowledge.

Significant Discoveries and Case Studies

Throughout the history of cosmic archaeology, there have been several monumental discoveries and fascinating case studies that have significantly shaped our understanding of extraterrestrial influences on Earth and beyond. One such remarkable discovery is the unearthed remnants of an ancient alien artifact in a remote location on Earth, providing compelling evidence of past extraterrestrial visitation. The meticulous analysis of this artifact, conducted using advanced scientific techniques, revealed intricate markings and materials unlike anything found on Earth, challenging conventional archaeological paradigms and igniting intense scientific interest worldwide. This unprecedented finding not only fundamentally challenged existing theories but also underscored the critical importance of cosmic archaeology in unraveling enigmatic aspects of human history. Another extraordinary case study surrounds the unearthing of an ancient structure with perplexing geometric patterns that align with celestial configurations, suggesting deliberate celestial alignment by a highly advanced ancient civilization. These profound revelations have sparked vigorous debates among archaeologists, historians, and astronomers, prompting collaborative efforts to decipher the cosmic significance of these enigmatic structures. The investigation of cosmic material traces, including extraterrestrial isotopic anomalies and nano-scale elements, has also yielded remarkable insights into the early interaction between our planet and cosmic visitors. By scrutinizing these traces through state-of-the-art analytical methodologies, researchers have unveiled compelling evidence of extraterrestrial intervention in key evolutionary periods of Earth's history—transforming our comprehension of humanity's cosmic connections. Moreover, the detailed examination of ancient artifacts such as intricately carved sculptures and enigmatic relics has unveiled striking parallels with iconic cosmic phenomena, shedding light on the influence of celestial events on ancient cultures and civilizations. The implications of these findings transcend conventional archaeological narratives, accentuating the indispensable role of cosmic archaeology in enriching our understanding of human history within the broader context of the cosmos. These exceptional cases represent just a fraction of the groundbreaking discoveries and thought-provoking case studies that continue to redefine the boundaries of cosmic archaeology, fostering a profound appreciation for the enduring interplay between terrestrial and extraterrestrial realms.

Challenges and Limitations

As with any scientific endeavor, cosmic archaeology is not without its challenges and limitations. The intersection of cosmology and archaeology presents a unique set of obstacles that researchers must navigate in order to advance the field. One of the primary challenges lies in deciphering the intricacies of extraterrestrial artifacts and evidence, which may be vastly different from terrestrial archaeological finds. The lack of a standard framework for interpreting cosmic relics further complicates the analysis and understanding of these ancient objects. Additionally, the sheer vastness of space presents logistical and practical hurdles. Accessing potential extraterrestrial sites and artifacts, whether on celestial bodies or drifting in deep space, poses significant challenges in terms of exploration and retrieval. Technological limitations also play a crucial role in constraining the scope of cosmic archaeological research. Current instrumentation and tools may not always be suited for the delicate and complex tasks required in the study of ancient extraterrestrial remnants. Furthermore, the interdisciplinary nature of cosmic archaeology necessitates collaboration across various scientific domains, which can present both logistical and communication challenges. Coordinating efforts between astrophysicists, archaeologists, geologists, and other experts requires effective communication and a shared understanding of each discipline's methodologies and terminology. Moreover, funding and resources for cosmic archaeological expeditions and studies are often limited compared to other scientific disciplines, impeding the pace of progress in this burgeoning field. Ethical considerations also pose a challenge in cosmic archaeology. As researchers delve into potential extraterrestrial civilizations and their artifacts, they must grapple with moral and ethical dilemmas related to the preservation and interpretation of alien relics. These challenges, however, serve as opportunities for growth and innovation within the field of cosmic archaeology. By addressing these obstacles head-on and fostering interdisciplinary collaboration, researchers can pave the way for groundbreaking discoveries and advancements in our understanding of ancient cosmic phenomena.

Collaboration Across Scientific Disciplines

Collaboration across scientific disciplines is at the heart of advancing the field of cosmic archaeology. This interdisciplinary approach brings together experts from diverse fields such as astronomy, geology, anthropology, physics, and biology to collectively address the complex challenges posed by cosmic archaeological research. The collaborative nature of this endeavor enables the pooling of knowledge, methodologies, and technologies from multiple disciplines, leading to enriched insights and holistic understanding.

Astronomy plays a pivotal role in providing cosmic context and celestial data crucial for identifying potential cosmic archaeological sites. Astrophysicists collaborate with cosmic archaeologists to analyze astronomical phenomena, such as supernovae or pulsar signals, which could indicate extraterrestrial activities. By integrating astronomical observations and archaeological findings, researchers can uncover patterns and correlations that offer valuable insights into the origins and interactions of extraterrestrial civilizations.

Geological expertise contributes significantly to the identification and dating of cosmic artifacts. Geologists work alongside cosmic archaeologists to analyze the geological composition of unearthed relics and determine their age and potential celestial origin. Understanding the geological contexts of cosmic artifacts provides essential clues about their provenance and the environments in which they may have existed, shedding light on the interplay between cosmic events and terrestrial landscapes.

Anthropological and sociological perspectives are crucial for interpreting potential cultural aspects of extraterrestrial civilizations. Collaboration with anthropologists allows cosmic archaeologists to apply anthropological frameworks to decipher extraterrestrial social structures, belief systems, and interactions. Insights from anthropology aid in contextualizing cosmic artifacts within the broader socio-cultural landscape, unraveling narratives of possible extraterrestrial societies and their contributions to cosmic heritage.

Physicists bring advanced analytical tools and theoretical models to cosmic archaeological investigations, enhancing the understanding of extraterrestrial technological capabilities and physical phenomena associated with cosmic relics. Their expertise in analyzing energy signatures, material properties, and quantum phenomena opens new frontiers in interpreting cosmic artifacts and discerning potential extraterrestrial interventions in natural processes.

Biology and astrobiology contribute valuable perspectives to cosmic archaeology by exploring the potential for extraterrestrial life forms and biological remnants within cosmic artifacts. Collaboration with biologists and astrobiologists enables the examination of organic components within unearthed relics, offering insights into hypothetical extraterrestrial biochemistry and indicating the presence of alien microbial life or genetic materials.

Overall, collaboration across scientific disciplines fosters a comprehensive and integrated approach to cosmic archaeology, transcending individual limitations and broadening the scope of inquiry. By embracing diverse expertise and perspectives, cosmic archaeologists can unravel the enigmatic tapestry of cosmic history and better comprehend our place within the cosmic ecosystem.

Technological Advancements Enhancing Cosmic Archaeology

Technological advancements have revolutionized the field of cosmic archaeology, enabling researchers to explore and analyze extraterrestrial artifacts and evidence with unprecedented precision and insight. One of the most notable technological advances is the development of advanced imaging techniques, such as high-resolution spectroscopy and multi-spectral imaging, which allow for the detailed examination of ancient relics and structures from distant cosmic origins. These cutting-edge imaging technologies provide invaluable data on the composition, structure, and history of celestial objects, shedding light on their potential cultural and scientific significance. Furthermore, the integration of artificial intelligence and machine learning algorithms has significantly enhanced the analysis of complex data sets, enabling researchers to identify patterns, anomalies, and potential correlations that may have previously gone undetected. The utilization of drones and remote sensing technologies has expanded the scope of cosmic archaeological exploration, facilitating the survey and mapping of extraterrestrial sites with unprecedented detail and accuracy. Moreover, advancements in robotics and autonomous exploration systems have enabled the deployment of sophisticated rovers and landers to explore challenging terrains on celestial bodies, retrieving invaluable samples and conducting on-site analyses. In addition, the development of state-of-the-art laboratory instrumentation and portable analytical devices has empowered researchers to conduct real-time analysis of extraterrestrial materials, allowing for rapid characterization and interpretation of unearthed relics. Furthermore, the emergence of virtual reality and augmented reality platforms has revolutionized the visualization and reconstruction of ancient cosmic environments, offering immersive experiences that transport researchers and enthusiasts to distant cosmic landscapes. The integration of these innovative technological tools not only enhances the efficiency and accuracy of cosmic archaeological research but also fuels public engagement and interest in the exploration of our cosmic heritage. As technological advancements continue to evolve, the future of cosmic archaeology holds immense promise, with the potential for groundbreaking discoveries and a deeper understanding of our place in the cosmos.

Future Perspectives in Cosmic Archaeology

The future of cosmic archaeology holds immense promise and potential, underpinned by the continuous evolution of technological innovations, interdisciplinary collaboration, and visionary thinking. As we peer into the horizon of cosmic archaeology, several key trends and perspectives emerge, shaping the trajectory of this dynamic field.

One of the most compelling future prospects revolves around the refinement and application of artificial intelligence (AI) and machine learning algorithms in cosmic archaeological research. AI, with its capacity for pattern recognition and data analysis, offers unprecedented opportunities to sift through vast datasets, recognize subtle extraterrestrial signatures, and unveil hidden connections across cosmic artifacts and phenomena. These advancements have the potential to revolutionize our understanding of ancient civilizations beyond Earth.

Furthermore, the convergence of cosmic archaeology with space exploration endeavors opens new vistas for discovery. With ambitious missions planned to explore celestial bodies within our solar system and beyond, including Mars, Europa, and Enceladus, there lies the prospect of unearthing traces of ancient extraterrestrial life or technology. The integration of cosmic archaeological methodologies with space missions presents an exciting frontier where tangible evidence of cosmic heritage may be uncovered, enriching our insights into the origins of life and civilizations.

Interdisciplinary collaboration stands as a linchpin for the future of cosmic archaeology. By fostering partnerships across diverse scientific domains - from astrophysics and astrobiology to anthropology and geology - an enriched tapestry of knowledge can be woven, transcending the boundaries of individual disciplines. This cross-pollination of expertise not only fuels innovative research but also cultivates a holistic approach to understanding the cosmic narrative, embracing its complexities and nuances.

Moreover, ethical considerations are poised to play an increasingly pivotal role in guiding the trajectory of cosmic archaeological pursuits. As the likelihood of encountering evidence of extraterrestrial civilizations grows, questions pertaining to cultural sensitivity, preservation of cosmic heritage, and responsible engagement with potential findings demand scrupulous attention. Examining these ethical dimensions proactively will ensure that the unfolding narrative of cosmic archaeology is steered by conscientious stewardship and respect for diverse cultures, both terrestrial and celestial.

In conclusion, the future of cosmic archaeology beckons with a tantalizing blend of technological innovation, interdisciplinary synergy, and ethical mindfulness. By embracing these perspectives and harnessing the collective ingenuity of the global scientific community, we are poised to unravel profound cosmic enigmas, shed light on our place in the universe, and perhaps encounter the enduring echoes of civilizations that once flourished amid the stars.

Chapter Four

The Cosmic Connection: Tracing Our Interstellar Origins

Interstellar Origins

Throughout history, humanity has sought to unravel the mysteries of its cosmic origins. The concept of interstellar heritage has sparked profound discussions among scholars and scientists, leading to groundbreaking explorations into the true nature of our connection to the cosmos. This inquiry into our interstellar origins holds significant implications for understanding the fundamental essence of human existence and the evolutionary pathways that have shaped our species. By delving into the paradigm of interstellar heritage, we can uncover the intricate threads that link our existence to the celestial realms beyond our planet. Through a comprehensive examination of historical narratives and scientific breakthroughs, we endeavor to elucidate the far-reaching influence of cosmic phenomena on human origins. As we embark on this intellectual journey, it becomes evident that our exploration of interstellar origins transcends conventional boundaries, transcending cultures, civilizations, and disciplines. The quest to comprehend our interstellar lineage is inherently interdisciplinary, encompassing astrophysics, cosmology, archaeology, anthropology, and astrobiology, weaving together a tapestry of knowledge that spans the depths of space and time. Exploring the concept of interstellar heritage necessitates a multidimensional approach that integrates empirical evidence, theoretical frameworks, and speculative imaginings. It invites us to contemplate the cosmic ballet that has choreographed the formation of our solar system and the genesis of life on Earth. Furthermore, this intellectual odyssey urges us to contemplate the role of cosmic events in shaping the course of human evolution and civilization, prompting us to reimagine our place in the grand narrative of the cosmos. As we delve deeper into the enigma of our interstellar origins, we are compelled to embrace a perspective that transcends terrestrial perspectives, expanding our consciousness to encompass the vast interstellar expanse that has sculpted the trajectory of human history. Unlocking the secrets of our interstellar heritage offers us a profound glimpse of our collective identity as stardust beings, bound by an intrinsic kinship with the universe itself.

Historical Perspectives on Cosmic Influence

Throughout human history, the notion of cosmic influence has been deeply embedded in various cultures and belief systems. From the ancient civilizations to the modern era, humans have constantly sought to understand their place within the cosmos and how celestial phenomena may have shaped our existence. The earliest records from civilizations such as the Sumerians, Egyptians, and Mayans reveal a profound reverence for celestial bodies and an understanding of their potential influence on earthly events. These ancient cultures often attributed natural phenomena, such as eclipses or planetary movements, to the whims of gods or higher beings, illustrating the early recognition of a connection between the celestial and terrestrial realms.

As scientific inquiry advanced, early astronomers such as Ptolemy, Copernicus, and Galileo laid the groundwork for understanding the mechanics of our solar system and the broader universe. Their observations and theories revolutionized humanity's conceptualization of our place in the cosmos, leading to paradigm shifts in cosmology and astronomy. Theories emerged that postulated Earth as just one of many celestial bodies orbiting the Sun, while others speculated about the possibility of countless other worlds beyond our own. These diverse perspectives gave rise to a deeper appreciation of the interconnectedness of celestial phenomena and their potential impact on terrestrial affairs.

In more recent centuries, technological advancements allowed for more sophisticated observations and measurements, leading to the birth of disciplines such as astrophysics and astrobiology. The discoveries made through these fields further illuminated the profound historical and ongoing influence of cosmic factors on Earth and its inhabitants. Evidence of extraterrestrial material, such as meteorites containing amino acids and other organic compounds, suggests a direct cosmic link to the origins of life on our planet. This intriguing connection has prompted scientists and researchers to delve deeper into the historical context of cosmic influence and its implications for the human story.

Moreover, the study of ancient myths, legends, and religious texts from cultures around the world reveals recurring themes of celestial beings and cosmic events shaping the course of human history. These narratives offer valuable insights into the enduring human fascination with cosmic influence and the enduring impact of such beliefs on cultural, social, and spiritual development. Understanding the historical perspectives on cosmic influence is crucial for appreciating the multifaceted nature of our interstellar origins and the profound implications that cosmic connections hold for our past, present, and future.

Methods of Tracing Cosmic Lineages

Tracing cosmic lineages presents a profound challenge that necessitates an interdisciplinary approach drawing from cosmology, archaeology, astrophysics, and planetary science. This pursuit requires the employment of sophisticated methodologies to discern and comprehend the intricate interplay between celestial phenomena and the evolution of terrestrial entities. A key aspect of tracing cosmic lineages involves the analysis of isotopic ratios in extraterrestrial materials such as meteorites, which can yield crucial insights into the origin and transformation of cosmic matter. Additionally, the field of comparative planetology provides an invaluable framework for understanding the diversity of planetary systems in our universe. By examining the structural composition, atmospheric characteristics, and geological features of other celestial bodies, scientists can elucidate the underlying processes that contribute to the formation and differentiation of planets, including Earth. Moreover, the study of stellar nucleosynthesis and the distribution of chemical elements throughout the cosmos offers a window into the fundamental building blocks that have shaped the development of planetary systems. Enhanced spectroscopic techniques enable the precise measurement and characterization of these elemental compositions, thereby allowing researchers to discern the signatures of various stellar sources imprinted within terrestrial materials. Furthermore, the investigation of galactic dynamics and the cosmological framework within which our solar system resides serves as a crucial foundation for comprehending the broader contextual influences that have sculpted the cosmic landscape. Advancements in computational modeling and simulation have revolutionized our capacity to simulate complex astrophysical events and their potential ramifications for planetary environments. Utilizing cutting-edge data analysis and visualization tools, researchers can unravel the dynamic interplay of cosmic forces that have left indelible imprints on the evolutionary trajectory of celestial bodies, including Earth. Integrating these diverse methodologies offers a comprehensive and nuanced approach to tracing cosmic lineages and elucidating the intricate narrative of our interstellar origins.

Astrophysical Dynamics and Earth's Formation

The formation of planet Earth is intimately linked to the larger landscape of astrophysical dynamics within the cosmos. Understanding the processes that led to the birth of our planet requires a deep dive into the complex interplay of celestial bodies, gravitational forces, and cosmic phenomena. As we venture into this exploration, we encounter the awe-inspiring narrative of cosmic evolution that set the stage for our planet's emergence. From the primordial chaos of stellar nurseries to the gravitational dance of protoplanetary disks, each step in this cosmic odyssey bears profound significance for Earth's eventual composition and characteristics. In tracing the intricate dance of celestial bodies, we unravel the story of Earth's formation—an epic tale scripted across the celestial canvas of space and time. The celestial ballet of supernovae, pivotal in the dispersal of heavy elements essential for terrestrial life, forms a crucial backdrop to Earth's cosmic lineage. Moreover, the cataclysmic collisions and gravitational perturbations within the early solar system sculpted the landscape on which life eventually took root. Delving deeper, we peer into the mesmerizing symphony of planetary accretion—a celestial artistry that culminated in the birth of Earth amidst the cosmic ballet. The dynamic interplay of gravitational perturbations, resonances, and tidal forces during this epoch not only shaped the physical geography of our planet but also laid the foundations for the existence of life-sustaining environments. Amidst the cosmic tapestry, Earth's position within the habitable zone emerged as a defining consequence of astrophysical dynamics. The fortuitous interplay of stellar luminosity, planetary orbits, and atmospheric evolution within this orbital haven set the stage for the flourishing of life on our planet. Unraveling the astrophysical intricacies that guided Earth's formation presents an enthralling voyage through the annals of cosmic evolution, illuminating the subtle interplay of celestial forces that birthed the cradle of terrestrial life.

Chemical Signatures of Stellar Contributions to Earth

As we delve into the realm of cosmic archaeology, the investigation of chemical signatures left by stellar contributions to Earth provides a fascinating window into our interstellar origins. Through meticulous analysis of isotopic ratios and elemental compositions, researchers have unearthed compelling evidence that points to the infusion of extraterrestrial material into the terrestrial landscape. The presence of rare isotopes such as xenon-129 and iridium, which are scarce on Earth but abundant in certain types of meteorites, accentuates the tangible link between celestial bodies and our planet.

By scrutinizing these chemical imprints, scientists endeavor to unravel the complex narrative of our solar system's formation and the profound role played by cosmic phenomena. The isotopic anomalies detected in ancient rocks and sediments serve as time capsules, preserving the fingerprints of distant stars and primordial cosmic events. These revelations challenge conventional perceptions of Earth's development, prompting us to contemplate the intimate interconnectedness of our world with the far reaches of space.

Moreover, the study of chemical signatures also extends to isotopic variations in the composition of noble gases and other elements, providing invaluable insights into the evolutionary history of our planet. The arrival of extraterrestrial materials, borne by comets, asteroids, and interstellar dust, has bestowed upon Earth a rich tapestry of elements that illuminate the intricate dance of cosmic forces. Through rigorous analytical techniques and cutting-edge instrumentation, the discernment of these exotic isotopic compositions has ushered in a new epoch of understanding where Earth's story is inexorably entwined with the broader cosmic narrative.

Furthermore, the convergence of interdisciplinary approaches, encompassing astronomy, geochemistry, and cosmochemistry, propels the elucidation of how the elemental abundances on Earth reflect a larger cosmic context. By examining the isotopic ratios of elements such as osmium, platinum, and palladium, researchers unlock the enigmatic chronicles of cosmic bombardment and the enduring legacy it has imparted to our planetary crucible. This engenders a profound appreciation for the harmonious interplay between the myriad constituents of space and the intricate choreography of our homeworld's formative epochs.

In conclusion, the exploration of chemical imprints derived from stellar sources underscores the remarkable confluence of celestial phenomena and terrestrial evolution. It redefines our perception of Earth as not just a product of local processes, but as an entity enriched by the cosmic mosaics woven across unimaginable spans of space and time. The saga of stellar contributions to Earth, imprinted in its very fabric, stands as a testament to the enduring interstellar heritage that shapes our existence.

Analyzing Isotopic Ratios: Clues to Alien Interventions

Isotopic ratios, the relative abundance of stable isotopes of elements, provide crucial insights into the history of cosmic interactions with Earth. By examining these isotopic compositions, scientists can discern potential anomalies that may signify extraterrestrial influence on our planet's development. It is within these minute differences in isotopic ratios that we seek evidence of alien interventions in Earth's geological and biological history. The analysis of isotopic ratios encompasses a diverse range of elements, each offering its own story of cosmic provenance. Oxygen isotopes, for instance, can reveal the source of water on Earth and shed light on the mechanisms behind the delivery of water from space. Similarly, the isotopic composition of certain elements in terrestrial rocks, such as iridium, can serve as an indicator of extraterrestrial impacts, providing critical evidence for cosmic interactions in Earth's ancient past. Beyond Earth's geology, isotopic analyses also yield invaluable insights into the development of life. The carbon-13 to carbon-12 ratio, for example, serves as a fingerprint for differentiating between biogenic and abiotic organic compounds. Moreover, isotopic fractionation processes can offer clues to the origins of complex organic molecules, hinting at potential celestial sources for the building blocks of life. Furthermore, the isotopic signatures found in the DNA of living organisms can provide evidence of genetic exchange with extraterrestrial sources, shedding light on the possibility of panspermia—the distribution of life across space via comets, asteroids, or interstellar dust. As we delve deeper into the realm of isotopic analysis, we stand at the precipice of uncovering profound revelations about the interconnectedness of Earth with the cosmos. These investigations not only enrich our understanding of Earth's history but also hold the potential to reshape our perceptions of humanity's place in the universe. Through rigorous scientific inquiry and the meticulous examination of isotopic ratios, we aim to discern the cosmic threads woven into the fabric of our existence, reaching beyond the confines of our planet to unravel the secrets of our interstellar origins.

Role of Comets and Asteroids in Seeding Earth

Comets and asteroids have long captured the imagination of scientists and enthusiasts alike. These celestial bodies, often referred to as the remnants of the early solar system, hold profound significance in the cosmic narrative of Earth's origins. The role of comets and asteroids in seeding Earth is a subject of great interest, and its implications reverberate across disciplines such as astronomy, planetary science, and astrobiology.

The influx of comets and asteroids during the early history of the Earth played a crucial role in shaping the planet's environment and delivering essential compounds. As these celestial bodies traversed the cosmos, they carried within them a diverse array of organic molecules, water, and other crucial building blocks of life. Upon impacting the young Earth, these materials contributed to the primordial soup that facilitated the formation of the first living organisms. The interplay between cometary and asteroidal impacts and the emergence of life on Earth showcases the profound influence of these cosmic visitors on our planet's evolutionary trajectory.

Furthermore, comets and asteroids are considered potential agents of panspermia, a hypothesis suggesting that life may have migrated between celestial bodies, potentially seeding different planets with the ingredients necessary for life. This concept prompts comprehensive investigations into the survivability of microorganisms within cometary or asteroidal environments, shedding light on the possibility of interplanetary transfer of life. The remarkable resilience of extremophiles, organisms capable of surviving in extreme conditions encountered in space, fuels speculation about the interstellar dispersal of life forms via these cosmic messengers.

The study of comets and asteroids also provides invaluable insights into the overall habitability of celestial bodies within and beyond our solar system. By examining the diversity of materials hosted by comets and asteroids, researchers gain essential data for understanding the potential habitable zones in distant planetary systems. Investigations into these celestial bodies contribute to the ongoing quest for identifying planets with conditions conducive to hosting extraterrestrial life. By uncovering the role of comets and asteroids in the seeding of Earth and evaluating their significance in the broader context of cosmic habitability, we take significant strides toward comprehending our interstellar origins.

Comparative Planetology and Habitability

As we delve into the intricacies of cosmic archaeology, it is essential to explore the concept of comparative planetology and its profound implications for understanding habitability across celestial bodies. The study of exoplanets, planetary diversity, and the conditions necessary to support life offers invaluable insights into the potential for extraterrestrial civilizations and the origins of life within our universe. A fundamental aspect of comparative planetology involves examining the physical characteristics of various celestial bodies, including their size, composition, atmosphere, and proximity to their host stars. Through this comparative analysis, scientists can discern the critical factors that contribute to a planet's potential to sustain life. Moreover, the interdisciplinary nature of comparative planetology integrates concepts from astronomy, geology, atmospheric science, and biology to distinguish habitable environments from hostile ones. This integrative approach facilitates comprehensive assessments of planetary conditions, enabling researchers to identify the underlying mechanisms that govern habitability. Furthermore, the comparative analysis of exoplanetary systems unveils the astonishing diversity of planetary compositions, climates, and potential habitats beyond our solar system. By leveraging advanced observational techniques and computational models, astronomers can discern the intricate interplay between a planet's geological history, atmospheric evolution, and ability to maintain suitable conditions for life. Notably, comparative planetology also underscores the significance of Earth as a model for habitability assessment, allowing researchers to contrast our planet's biosphere with diverse environmental extremes found elsewhere in the cosmos. This juxtaposition provides valuable benchmarks for evaluating the plausibility of life on other worlds and the resilience of organisms in different planetary contexts. Additionally, the pursuit of comparative planetology amplifies our comprehension of the delicate equilibrium required to sustain life, broadening our understanding of the unique circumstances that enabled the proliferation of life on Earth. As we venture into this enthralling realm of knowledge, the quest for understanding habitability transcends conventional boundaries, instilling a sense of wonder and curiosity about the myriad possibilities existing beyond our terrestrial abode.

Impact of Galactic Events on Earth's Biology

The impact of galactic events on Earth's biology has been a subject of great intrigue and exploration within the field of cosmic archaeology. While Earth's immediate environment plays a crucial role in shaping its biological evolution, the influence of galactic phenomena cannot be discounted. Over the expanse of geological time, the Earth has been subject to various extragalactic events, ranging from cosmic ray bursts to interstellar dust influx, each leaving an indelible mark on the trajectory of life on our planet.

One significant aspect is the occurrence of supernovae within close proximity to the solar system. These cataclysmic events unleash a torrent of high-energy particles and radiation, which can have profound effects on Earth's biosphere. The increased flux of cosmic rays has the potential to induce mutations in terrestrial organisms, thereby impacting evolutionary processes. Moreover, the deposition of isotopic anomalies resulting from supernovae ejecta provides tangible evidence of galactic influences in Earth's geological archives.

Another galactic phenomenon with repercussions for Earth's biology is the periodicity of comet and asteroid impacts driven by interactions within the solar system and the Milky Way galaxy. Galactic gravitational perturbations can lead to an enhanced influx of cosmic debris, thereby exposing the Earth to heightened impact events. These extraterrestrial collisions have been implicated in mass extinction events and the restructuring of ecosystems, highlighting the interconnectedness of galactic dynamics and terrestrial life.

Furthermore, fluctuations in galactic environments, such as variations in cosmic ray intensity and the prevalence of background radiation, can exert selective pressures on the development and sustenance of life on Earth. The intricate interplay between Earth's magnetosphere and galactic cosmic rays intricately influences atmospheric chemistry, climate, and ultimately biological processes. Researchers continue to investigate the extent of these connections and their role in shaping the course of evolution on Earth.

Understanding the impact of galactic events on Earth's biology not only sheds light on the past but also holds implications for the future. As humanity ventures into space exploration and colonization, comprehending the broader galactic context becomes increasingly significant. Unraveling the intricate relationship between Earth and the cosmos presents an opportunity to comprehend the factors that have steered biological evolution and to anticipate the challenges and opportunities that await as we extend our presence beyond our home planet.

Summary & Implications for Future Research

The preceding discussion delved into the profound impact of galactic events on Earth's biology, shedding light on the intricate interplay between cosmic phenomena and the development of life on our planet. As we examine the implications of this cosmic connection, it becomes evident that our understanding of the universe holds invaluable insights for future research in various fields. Firstly, the recognition of galactic events as influential factors in Earth's biological evolution prompts a reevaluation of astrobiological principles. By integrating findings from cosmology, planetary science, and biology, scientists are poised to formulate more comprehensive theories regarding the origins of life and the potential distribution of life beyond Earth. The interdisciplinary approach is paramount in elucidating the broader significance of cosmic influences on habitability and the emergence of complex organisms.

Chapter Five

Patterns in Time: From Earth's Origins to the Cosmos

The Timeline of the Universe

The timeline of the universe is a complex tapestry of events that stretches back over 13.8 billion years. At its very beginning was the Big Bang, an event of unfathomable energy and density from which the universe as we know it emerged. In the immediate aftermath of the Big Bang, the universe underwent a rapid expansion known as cosmic inflation, during which time fundamental forces and particles came into existence. Over billions of years, this expanding cosmos began to cool and allow the formation of subatomic particles, ultimately leading to the creation of protons, neutrons, and electrons. These particles coalesced to form the first atoms, primarily hydrogen and helium. Gradually, these atoms congregated into vast clouds, laying the groundwork for the formation of the first stars and galaxies. As these early celestial entities ignited, they set off a profound epoch of stellar evolution and elemental synthesis. The birth and death of stars churned out heavier elements, which were dispersed throughout the universe in supernova explosions, enriching the interstellar medium with the raw materials necessary for planetary systems and life. Our own solar system, including the Earth, formed from one such cloud of gas and dust roughly 4.6 billion years ago. This brief summary scratches only the surface of the immense tapestry of cosmic history, yet it highlights the grand orchestration of events that have led to the emergence of our planet and the life it supports. Understanding the timeline of the universe is crucial not only for unraveling the origins of Earth but also for comprehending our place in the cosmic tableau.

Geological Clocks: Earth's Formative Years

The geological clocks of Earth provide a fascinating look into the formative years of our planet's history. By analyzing the layered records of rocks and minerals, scientists have been able to decipher the intricate timeline of geological events that shaped our world. Through the methods of radiometric dating, such as uranium-lead dating and potassium-argon dating, researchers have gained valuable insights into the age of various geological formations, allowing them to unravel the sequence of Earth's transformative stages. These findings not only shed light on the chronology of geological processes but also provide a framework for understanding the interplay between cosmic phenomena and their impact on Earth.

Moreover, isotopic dating techniques have enabled scientists to discern the ages of ancient rocks and uncover crucial information about the formation of continents, volcanic activity, and evolutionary changes in Earth's crust. This has been pivotal in constructing a comprehensive narrative of the early Earth, including the emergence of land masses, the onset of tectonic movements, and the evolution of life forms. The meticulous analysis of geological time scales has revealed how Earth transitioned from a primordial state to the diverse and vibrant planet we inhabit today.

Furthermore, the study of sedimentary rocks offers a unique perspective on Earth's environmental history. By examining sediment layers, researchers can glean insights into past climates, oceanic conditions, and ecological shifts. This is instrumental in understanding the dynamic interplay between terrestrial and celestial events, as well as their enduring impact on Earth's ecosystems. Additionally, the field of stratigraphy allows for the correlation of geological layers across different regions, facilitating the construction of a unified global timescale that traces the continuum of Earth's development.

In essence, the investigation of geological clocks provides a compelling narrative of Earth's formative years, offering a window into the profound interactions between cosmic processes and the evolution of our planet. The revelations garnered from these studies not only deepen our understanding of Earth's history but also serve as a testament to the enduring influence of the cosmos on our terrestrial home.

Astrophysical Processes and Their Significance on Earth

From the grand explosion of stars to the intricate dance of celestial bodies, astrophysical processes wield a remarkable influence on the development and sustenance of Earth. The birth and death of stars not only produce elements that form the building blocks of our planet but also set the stage for the emergence of life. Supernova explosions scatter these essential elements, such as carbon, nitrogen, and oxygen, across the cosmos, eventually coalescing into new stars, planetary systems, and habitable worlds. This cosmic orchestration of matter is fundamental in shaping the conditions necessary for the evolution of complex life forms on Earth. Moreover, as our solar system journeys through the Milky Way, it encounters various interstellar phenomena that leave indelible imprints on our planet's climate and geological dynamics. For instance, cosmic rays from distant supernovae, along with other interstellar particles, may have played a role in the modulation of Earth's climate and biological evolution. Furthermore, gravitational interactions with neighboring celestial bodies influence Earth's axial tilt and orbit, contributing to long-term climatic oscillations and mass extinction events. Studying these astrophysical processes provides invaluable insights into the deep-time history of our planet and the intricate web of connections between the cosmos and Earth. As humanity ventures beyond the confines of our home world, unraveling the multifaceted relationships between astrophysical phenomena and Earth will be pivotal in understanding and preserving the delicate equilibrium that sustains life on our unique celestial oasis.

Celestial Phenomena and Earth's Development

Throughout the history of our planet, celestial phenomena have played a profound role in shaping Earth's development. From the gravitational influence of the moon on tides to the impact of asteroid strikes, the interaction between cosmic events and terrestrial processes has left an indelible mark on the planet's evolution. The cyclical nature of celestial phenomena, such as the shifting axial tilt and orbital eccentricity, has contributed to periods of climatic change that have directly impacted the course of life on Earth. The study of celestial phenomena and their correlation with terrestrial dynamics provides invaluable insights into the intricate interconnectedness of the cosmos and our home planet.

One of the most significant celestial phenomena impacting Earth is the phenomenon of solar radiation. The energy emitted by the sun not only sustains life on our planet but also influences its climate and atmospheric conditions. Variations in solar output have been linked to climatic fluctuations and have shaped the environmental conditions necessary for the emergence and evolution of life. Understanding the complex interplay between solar radiation and Earth's systems is crucial for comprehending the long-term patterns of Earth's development.

Another celestial event that has profoundly influenced Earth is the occurrence of cosmic collisions. Impact events, such as asteroid strikes, have had cataclysmic effects on the planet, leading to mass extinctions and altering the trajectory of biological evolution. These cosmic collisions have left enduring signatures in the geological record, providing valuable clues about the tumultuous history of our world. By studying the remnants of these impacts, scientists can glean insights into the frequency and magnitude of such events and their far-reaching consequences for Earth.

Moreover, the interactions between Earth and cosmic phenomena extend beyond direct physical effects. Cultural and societal developments have been influenced by celestial events, leading to the incorporation of cosmic observations into human belief systems, calendars, and rituals. The recognition of celestial cycles and their alignment with key cultural milestones underscore the pervasive impact of cosmic phenomena on human civilization, underscoring the intrinsically intertwined nature of humanity and the cosmos.

In conclusion, the exploration of celestial phenomena and their relationship with Earth's development offers a holistic perspective on the interconnectedness of the universe. By delving into the intricate dance between cosmic events and terrestrial processes, we deepen our understanding of the forces that have sculpted our planet over billions of years, illuminating the profound bond between Earth and the wider cosmos.

Synchronization of Cosmic and Earth Events

The synchronization of cosmic and Earth events is a captivating subject that unveils the intricate dance between celestial phenomena and the development of our planet. Throughout history, astronomers, geologists, and archaeologists have sought to unravel the profound connections between events in the cosmos and their impact on Earth. From astronomical alignments shaping ancient monuments to cosmic influences on climate patterns, the concept of synchronization encapsulates a wide array of phenomena. At the heart of this exploration lies the recognition that cosmic events do not occur in isolation; they are intricately linked to the processes unfolding on our planet. This interconnectedness forms the basis for understanding the dynamic relationship between celestial occurrences and terrestrial evolution. A key aspect of this synchronization is the interaction between celestial bodies and Earth's geophysical systems. The gravitational influence of the moon, for instance, manifests in the ebb and flow of tides, affecting coastal landscapes and ecosystems. Beyond the moon, planetary alignments and solar cycles exert subtle but impactful influences on Earth's climatic and geological patterns. Exploring the historical records left by ancient civilizations, we can discern how celestial events were meticulously observed and integrated into cultural and societal practices. Astronomical alignments manifest in the orientation of architectural structures such as Stonehenge, demonstrating a deep awareness of cosmic rhythms. Moreover, the study of cosmic-ray exposure dating and the identification of extraterrestrial components in geological strata offer intriguing insights into the synchronization between cosmic events and terrestrial phenomena. These remarkable correlations provide a window into the deep-time dynamics that have shaped our planet. As we delve further into this fascinating realm, it becomes evident that the synchronization of cosmic and Earth events is not merely a historical curiosity, but a fundamental aspect of our ongoing ecological and geological narrative. By delving into the geological and astronomical archives, researchers continue to unveil new layers of evidence that highlight the intricate nature of this connection. Understanding the synchronization of cosmic and Earth events enriches our comprehension of Earth's history and enhances our ability to anticipate potential future changes. Exploring this relationship provides a holistic perspective, underscoring the remarkable interplay between Earth and the cosmos.

Radiometric Dating and Cosmic Chronology

Radiometric dating plays a crucial role in unraveling the mysteries of cosmic chronology, allowing us to trace the timeline of celestial events and their profound impact on Earth's development. By examining the abundance of isotopes within rocks and meteorites, scientists can determine the ages of these materials with exceptional precision. This method enables us to reconstruct the sequence of cosmic phenomena, including the formation of stars, the birth of galaxies, and the elemental composition of our universe. Furthermore, radiometric dating provides valuable insights into the timing of key events in Earth's history, such as the formation of the primordial crust and the emergence of life. Through the analysis of radioactive decay and the subsequent transformation of isotopes, researchers are able to establish a comprehensive chronology of the cosmos, shedding light on the intricate interplay between celestial processes and terrestrial evolution. As we delve deeper into the realms of radiometric dating, we gain a profound understanding of the dynamic relationship between cosmic events and their enduring repercussions on our planet. By meticulously analyzing the isotopic signatures embedded within ancient rocks and cosmic debris, we gain unprecedented access to the profound history of our universe and the enigmatic forces that have shaped the course of cosmic evolution. It is through this meticulous investigation that we can decipher the timeline of celestial phenomena, from the explosive birth of stars to the cataclysmic collisions that have sculpted the fabric of space and time. Radiometric dating serves as an indispensable tool in unlocking the secrets of cosmic chronology, presenting us with a captivating narrative of the cosmic ballet that has unfurled over billions of years. The insights gleaned from this method transcend the confines of terrestrial boundaries, offering a panoramic view of the cosmic drama that has unfolded across the vast expanse of the universe. With each layer of cosmic chronology unveiled through radiometric dating, we inch closer to comprehending the intricate tapestry of the cosmos and the indelible imprints left by celestial phenomena on our planet.

Impact of Meteoritic Events on Earth's Biological Evolution

Meteoritic events have played a pivotal role in shaping the course of Earth's biological evolution. The impact of large meteorites and asteroids has caused catastrophic environmental changes, leading to mass extinctions and significant shifts in biodiversity over millions of years. These extraterrestrial collisions have left profound imprints on Earth's geological and biological history, with some of the most notable impacts including the Chicxulub impact event believed to be associated with the extinction of non-avian dinosaurs. Beyond the catastrophic effects, meteoritic events have also played a paradoxical role in evolutionary processes. Following mass extinctions, the subsequent ecological vacuums created opportunities for new species to emerge and flourish, driving evolutionary innovation. Evidence suggests that certain meteoritic materials may have contributed crucial elements necessary for the emergence and sustenance of life on Earth, potentially seeding the planet with organic compounds necessary for the origin of life. Additionally, the flux of meteoritic material entering the Earth's atmosphere provides a continuous source of genetic material, influencing the genetic diversity and adaptive potential of terrestrial organisms. Furthermore, meteorite impacts have influenced the development of planetary environments, affecting climate, atmospheric composition, and oceanic conditions, which in turn have had profound impacts on the evolution and distribution of life on Earth. The study of meteoritic events and their consequences continues to provide valuable insights into the interconnectedness of terrestrial and cosmic phenomena and has significant implications for our understanding of Earth's historical and future trajectory.

The Role of Solar Activity in Terrestrial Climate

Solar activity plays a crucial role in shaping the Earth's climate, exerting a profound influence on various aspects of our planet's environmental dynamics. The Sun's energy output is not constant and experiences fluctuations over different timescales, impacting the Earth's atmospheric and climatic conditions. Understanding the mechanisms through which solar activity affects terrestrial climate is essential for comprehending long-term climate patterns and making accurate projections for future climate changes.

One of the primary ways solar activity impacts the Earth's climate is through variations in solar radiation. Solar radiation, particularly in the form of sunlight, provides the energy that drives the Earth's climate system. Changes in solar irradiance, which refers to fluctuations in the amount of solar energy reaching the Earth, can lead to alterations in global temperature patterns and precipitation levels. Studying historical records and paleoclimate data allows researchers to discern correlations between solar activity and past climate shifts, providing valuable insights into the complex interactions between the Sun and Earth's climate.

Solar activity also influences the Earth's climate through its impact on the planet's upper atmosphere and ionosphere. Variations in solar ultraviolet (UV) radiation can lead to changes in the ionization and composition of the Earth's upper atmosphere, affecting atmospheric circulation patterns and influencing regional and global climate dynamics. Additionally, solar activity can modulate the influx of cosmic rays reaching the Earth, which may have implications for cloud formation processes and overall climate stability.

In addition to direct radiative effects, solar activity can indirectly influence terrestrial climate through its potential association with multi-decadal and centennial climate oscillations. Researchers have observed potential links between variations in solar activity and phenomena such as the solar cycle, the 11-year periodic variation in solar magnetic activity, and climate oscillations like the Atlantic Multidecadal Oscillation (AMO) and the Pacific Decadal Oscillation (PDO). These associations underscore the intricate interplay between solar dynamics and long-term climate variability, offering fertile ground for further exploration and analysis.

As we strive to unravel the complexities of Earth's climate system, ongoing research efforts continue to elucidate the multifaceted ways in which solar activity shapes terrestrial climate. By integrating knowledge from solar physics, atmospheric science, and paleoclimatology, scientists are working to refine our understanding of solar-climate connections and their implications for current and future climate trends. Ultimately, grasping the intricate relationship between solar activity and terrestrial climate is integral to advancing our ability to model and predict climatic changes, thereby facilitating informed decision-making and adaptive strategies in the face of evolving environmental conditions.

Interstellar Matter and Its Influence on Earth

The influence of interstellar matter on Earth is a topic of profound significance, delving into the link between celestial phenomena and terrestrial events. Interstellar matter, composed primarily of gas and dust, plays a pivotal role in shaping planetary systems, including our own. Its impact extends beyond the astronomical realm, permeating into the very fabric of Earth's history and development. The influx of interstellar material has left discernible imprints on our planet, contributing to a deeper understanding of cosmic interactions with the Earth.

One notable aspect of interstellar matter's influence is its role in the formation of celestial bodies. Within vast interstellar clouds, gravitational forces condense gas and dust particles, birthing new stars and planetary systems. As these systems evolve, they become intertwined with the cosmic environment, influenced by the remnants of their stellar natal surroundings. This process extends to our solar system, as Earth and other planets bear the signatures of their interstellar heritage.

Moreover, interstellar matter continually interacts with Earth through cosmic phenomena such as supernovae and cosmic rays. Supernova explosions disperse heavy elements forged within stars throughout the universe, enriching the interstellar medium and seeding nascent planetary systems with the building blocks of life. Likewise, cosmic rays, consisting of high-energy particles, penetrate the solar system from interstellar space, impacting Earth's atmosphere and climatic processes. These cosmic rays have been proposed to influence cloud formation and thereby exert a potential role in shaping Earth's climate.

Furthermore, the presence of interstellar matter in the vicinity of the solar system has direct implications for the study of exoplanets and astrobiology. As astronomers identify exoplanetary systems in various stages of formation, the imprint of interstellar matter becomes a crucial factor in understanding the diversity and composition of exoplanetary environments. Additionally, the potential transport of organic compounds within interstellar material raises intriguing questions about the origins of life on Earth and the potential for similar processes in extraterrestrial habitats.

In summary, the influence of interstellar matter on Earth encompasses a wide spectrum of phenomena, ranging from the formation of planetary systems to its potential roles in climatic and biological evolution. By examining these cosmic interactions, we gain invaluable insights into the interconnectedness of celestial and terrestrial realms, elucidating the profound influence of the cosmos on our home planet.

Cosmological Shifts and Their Implications for Earth

The study of cosmological shifts and their consequences for Earth is a multifaceted endeavor that delves into the depths of astrophysics and geology. Cosmic phenomena such as galactic interactions, cosmic rays, and changes in the solar system's orbit can have profound implications for our planet's climate and geological processes. Understanding these cosmological shifts and their effects on Earth is crucial for comprehending the broader tapestry of our planet's history and future.

One fundamental area of focus within this realm is the examination of periodicity in cosmic events and their potential impact on Earth. While terrestrial geological processes unfold over vast timescales, cosmic phenomena often operate on much grander temporal cycles. The correlation between cosmic events and geological phenomena suggests an intricate interplay that has left enduring imprints on Earth's historical narrative.

Moreover, cosmological shifts have been linked to mass extinction events in Earth's history. Investigating the triggers of these cataclysmic occurrences, such as asteroid impacts or volcanic eruptions, requires a holistic perspective that incorporates cosmic influences. By analyzing cosmological data alongside Earth's geological record, researchers can unravel the complex dynamics that have shaped the planet's evolutionary trajectory.

Equally significant is the examination of how cosmological shifts may influence Earth's climate patterns. Variations in solar activity, cosmic radiation levels, and celestial dynamics can steer climatic shifts on our planet. These insights are pivotal for interpreting past climatic trends and forecasting potential future changes. The implications of these cosmological shifts extend beyond mere academic interest, offering critical perspectives for addressing contemporary climate change challenges.

Furthermore, by investigating the interplay between cosmological phenomena and Earth's magnetic field, scientists can delve into the mechanisms driving geomagnetic reversals. These periods of magnetic upheaval have wielded substantial impact on Earth's environment and life forms, making them a subject of intense scrutiny within the context of cosmological shifts.

In essence, comprehending the ramifications of cosmological shifts for Earth necessitates a comprehensive fusion of astronomical, geological, and climatological disciplines. This concerted approach holds the key to elucidating the interconnectedness of cosmic events and terrestrial realities, presenting a compelling tapestry of cosmic influences on our planet's past, present, and future.

Chapter Six

Ancient Artifacts: Unraveling the Mysteries of Extraterrestrial Relics

Extraterrestrial Relics

Artifacts with potential extraterrestrial origins hold immense significance in the fields of archaeology, astronomy, and astrobiology. These relics, often discovered in ancient archaeological sites, present a rare opportunity to explore the possibility of extraterrestrial influence on Earth's history. The study of such artifacts not only offers a window into civilizations beyond our planet but also challenges conventional historical narratives. The rarity of these relics adds another layer of intrigue, as their identification and verification require rigorous methodologies and interdisciplinary collaboration. The implications of discovering an object whose origins lie beyond Earth are profound, shaping our understanding of both terrestrial and extraterrestrial histories. Furthermore, the investigation of these relics opens avenues for contemplating the technological advancements of potential alien civilizations and the complexities of interactions between different planetary systems. As we delve into the following sections, we aim to unravel the mysteries surrounding these artifacts and shed light on their potential implications for humanity's broader cosmic context.

Methodologies for Identifying Non-Terrestrial Origins

The quest to identify and understand relics of non-terrestrial origins demands a rigorous approach founded on robust methodologies. In the layered tapestry of extraterrestrial artifact analysis, scientific scrutiny is paramount. The foundation of this methodological framework is rooted in interdisciplinary collaboration, drawing from fields such as astrophysics, geochemistry, geology, archaeology, and materials science.

One of the primary methodologies is the utilization of advanced spectroscopic techniques, including X-ray fluorescence (XRF) and Raman spectroscopy. These methods enable researchers to analyze the elemental and molecular composition of extraterrestrial artifacts, providing critical insights into their origin and potential journey through space. Additionally, isotopic analysis plays a pivotal role in discerning whether the materials comprising these relics could have originated beyond Earth's confines.

Furthermore, the application of cutting-edge imaging technologies, such as scanning electron microscopy (SEM) and high-resolution computed tomography (CT), allows for a detailed examination of the intricate features and internal structures of the artifacts. This not only aids in identifying any unconventional or non-terrestrial manufacturing processes but also provides invaluable clues regarding the environmental conditions in which these artifacts may have been created.

Beyond the realm of material analysis, the study of extraterrestrial relics necessitates the incorporation of astronomical data and computational simulations to trace their potential celestial origins. Utilizing data from astronomical surveys and telescopic observations, researchers can explore plausible trajectories that these artifacts might have traversed before arriving on Earth. Coupled with computational modeling, this approach enables the reconstruction of possible impact events and planetary dynamics that could have propelled these relics across the cosmic expanse.

The coherent integration of these methodologies with meticulous fieldwork and archaeological excavation procedures is essential in establishing a comprehensive understanding of the provenance and context of extraterrestrial artifacts. This includes the systematic documentation of contextual information such as stratigraphy, spatial distribution patterns, and any co-associated indigenous cultural remains, which collectively contribute to elucidating the chronological and cultural significance of these enigmatic relics.

Unearthing the secrets shrouding extraterrestrial artifacts requires an unwavering commitment to precision, objectivity, and scientific rigor, underpinned by a resolute dedication to unraveling the mysteries that bind our cosmic narrative with potential off-world influences.

Notable Extraterrestrial Artifacts Discovered

Throughout history, numerous intriguing artifacts have been discovered that provoke questions about their extraterrestrial origins. One such artifact, the Antikythera mechanism, found in a shipwreck off the Greek island of Antikythera, is a prime example of advanced ancient technology. Dating back to the 2nd century BCE, this intricate device consists of a complex assembly of gears and cogs, leading researchers to speculate on its purpose and origin. The sophistication of this artifact has led some experts to consider the possibility of extraterrestrial influence or intervention in its creation.

Another fascinating case is that of the Dropa stones, allegedly discovered in a remote mountainous region of China. These circular discs feature a mysterious hieroglyphic script that eludes translation, and according to local legend, are linked to an extraterrestrial visitation thousands of years ago. The enigmatic nature of these artifacts challenges conventional archaeological explanations and continues to captivate the imagination of researchers and enthusiasts alike.

In the deserts of Peru, the Nazca lines present another compelling mystery. These massive geoglyphs, visible only from the air, depict various animals, plants, and geometric shapes. Some theorists propose that these designs could have been created with extraterrestrial input, suggesting a cosmic connection to their purpose and construction. The profound scale and precision of the Nazca lines raise thought-provoking questions about the involvement of advanced alien entities in ancient Earthly civilizations.

Moving to modern times, the Tulli Papyrus, an ancient Egyptian document dating back to the reign of Thutmose III, contains an account of strange flying objects witnessed by Egyptian scribes. This intriguing record raises inquiries into potential ancient encounters with advanced extraterrestrial craft, challenging established perceptions of technological capabilities in ancient cultures.

These cases represent only a fraction of the notable extraterrestrial artifacts discovered throughout history, each offering unique insights and provocative suppositions. While differing viewpoints exist regarding the origins of these artifacts, the exploration of such compelling cases stimulates discourse and drives forward the pursuit of understanding the potential cosmic influences on human history.

Material Composition and Analysis Techniques

The study of material composition and analysis techniques in the field of extraterrestrial relics is a multifaceted endeavor that requires a comprehensive approach to unravel the mysteries surrounding these artifacts. By applying advanced scientific methodologies, archaeologists and cosmologists can gain valuable insights into the origin, age, and nature of these enigmatic objects. Material composition analysis involves the identification and characterization of the constituent elements and compounds present in the relics. This process often entails the use of cutting-edge technologies such as spectroscopy, mass spectrometry, and X-ray fluorescence, which provide detailed information about the atomic and molecular makeup of the materials. These analytical techniques enable researchers to determine whether the materials have terrestrial or extraterrestrial origins, offering crucial evidence in the quest to understand the provenance of the artifacts. Additionally, isotopic analysis plays a pivotal role in ascertaining the isotopic ratios of certain elements present in the relics, providing invaluable data for discerning their cosmic or earthly genesis. Moreover, the examination of crystalline structures through methods like X-ray diffraction facilitates the identification of specific mineral phases and crystallographic properties—essential for establishing the unique characteristics of these extraordinary relics. Furthermore, the study of microstructures utilizing electron microscopy allows for the visualization of minute features within the materials, shedding light on any potential alien manufacturing processes or external influences. In tandem with material composition analysis, dating techniques are employed to determine the age of the artifacts. Radiometric dating methods, including carbon-14 dating and uranium-lead dating, are instrumental in establishing the temporal context of the relics, aiding in the formulation of timelines for extraterrestrial interactions with our planet. Optically stimulated luminescence dating also proves beneficial in assessing the last exposure of materials to sunlight, contributing to the chronology of the artifacts' arrival or deposition on Earth. The integration of these diverse analysis techniques not only elucidates the physical and chemical properties of the relics but also provides compelling evidence to support the notion of extraterrestrial influence. By applying rigorous scientific methodologies, experts continue to unravel the complexities of these extraordinary artifacts, paving the way for a deeper understanding of our cosmic origins.

Technological Implications of Alien Craftsmanship

The study of ancient artifacts of potential extraterrestrial origins presents a unique opportunity to analyze the technological advancements of non-terrestrial civilizations. Understanding the technological implications of alien craftsmanship involves a multidisciplinary approach that encompasses fields such as material science, engineering, and archaeology. By scrutinizing the design, construction, and functionality of these artifacts, researchers can gain insight into the advanced capabilities of extraterrestrial cultures. The analysis of materials utilized in these artifacts can offer valuable information about the level of knowledge and technology possessed by the creators. Advanced metallurgical techniques, nanotechnology, or materials not found on Earth could provide evidence of highly sophisticated manufacturing processes. Moreover, the mechanisms and intricate components within these relics may reveal engineering feats that surpass contemporary human capabilities. By comparing these findings with known terrestrial technologies, researchers can discern distinct differences and potentially identify innovations that are beyond our current understanding. Furthermore, the study of extraterrestrial artifacts allows for the exploration of unconventional energy sources, propulsion systems, and communication devices. Unearthed relics with unknown power sources or peculiar energy distribution patterns could offer glimpses into alternative methods for harnessing and utilizing energy. Additionally, the examination of propulsion systems and spacecraft components may shed light on technologies that enable interstellar travel or atmospheric manipulation. Such discoveries have the potential to revolutionize our understanding of physics and engineering, inspiring new avenues of research and technological development. As we uncover more about the technological implications of alien craftsmanship, it becomes increasingly evident that the artifacts hold the key to unlocking enigmatic mysteries and expanding the frontiers of human knowledge. By meticulously studying and scrutinizing these relics, we endeavor to bridge the gap between terrestrial and extraterrestrial technologies, providing invaluable insights into the capabilities of civilizations beyond our world.

Temporal Disparities: Dating the Artifacts

Dating extraterrestrial artifacts poses a unique challenge due to the potential temporal disparities between these objects and terrestrial civilizations. Traditional dating methods such as carbon dating and stratigraphy may not be sufficient to accurately determine the age of extraterrestrial relics. One approach involves examining isotopic compositions to reveal the decay rates of elements that differ from those found on Earth. This allows for the estimation of artifact ages, offering intriguing insights into ancient cosmic timelines. Another promising method is luminescence dating, which measures the accumulation of trapped electrons in crystal structures over time. This technique could significantly expand our understanding of extraterrestrial artifact chronology. Additionally, considering the potential technological advancements of alien civilizations, it becomes imperative to use interdisciplinary approaches in dating these artifacts. Collaborations with astrophysicists, geologists, and archaeologists can offer diverse perspectives in establishing the temporal context of extraterrestrial relics. Furthermore, the exploration of artifacts from various celestial bodies, such as meteorites or lunar samples, presents an opportunity to correlate their ages with known cosmic events. Understanding the temporal aspects of these artifacts not only sheds light on their origins but also provides critical information regarding the interactions between extraterrestrial civilizations and our own. In the quest to decipher the mysteries of ancient cosmic civilizations, the accurate dating of extraterrestrial artifacts stands as a fundamental pillar, unlocking pivotal chapters in the cosmic history of our universe.

Geographic Distribution Patterns of Finds

The study of extraterrestrial relics demands a comprehensive analysis of the geographic distribution patterns of these enigmatic finds. Understanding where these artifacts are located and their spatial relationships is crucial in unraveling the mysteries they hold. Geographic information systems (GIS) play a pivotal role in mapping and visualizing the distribution patterns of these relics. By employing GIS technology, researchers can discern potential hotspots of extraterrestrial artifact concentrations and investigate correlations with terrestrial ancient civilizations. Moreover, the examination of the geographical dispersion of these relics contributes to our understanding of potential interaction networks between different regions. Through GIS analysis, researchers can identify clusters or specific regions where these artifacts are prevalent, shedding light on possible patterns of extraterrestrial contact or influence. Furthermore, the geographic distribution patterns provide valuable insights into the shipping or trade routes that may have facilitated the transportation of these artifacts, offering clues about the extent of interstellar interactions in ancient times. Exploring the spatial relationships of these relics also aids in identifying areas rich in extraterrestrial remnants, guiding future archaeological expeditions and enhancing strategies for targeted field investigations. It is important to note that thorough analyses must consider environmental factors and geological contexts, as these variables can influence the preservation and distribution of ancient artifacts. Additionally, comparing the geographic distribution patterns of extraterrestrial relics with known terrestrial archaeological sites can reveal intriguing juxtapositions and potentially uncover interconnected narratives between human history and potential extraterrestrial influences. The careful examination of geographic distribution patterns, combined with other avenues of research, forms a multifaceted approach to decoding the complex tapestry of cosmic archaeology, ultimately contributing to a more holistic understanding of our place in the cosmos.

Theoretical Frameworks in Extraterrestrial Archaeology

The pursuit of understanding extraterrestrial artifacts and relics requires a solid theoretical framework to guide the investigation. This framework must be interdisciplinary, drawing on fields such as archaeology, anthropology, astrophysics, and exobiology. One key theoretical perspective is the idea that extraterrestrial artifacts can serve as tangible evidence of contact or interaction with other intelligent civilizations, both within our own star system and potentially beyond. This perspective motivates a rigorous approach to artifact analysis, seeking to discern any potential patterns or markers that may suggest intentional cultural or technological exchange. Another theoretical framework involves considering contemporary Earth's technology and cultural practices as a reference point for evaluating the functionality and purpose of unearthed extraterrestrial relics. Additionally, researchers must remain open to the possibility that these artifacts could represent phenomena or civilizations beyond our current understanding, necessitating a willingness to explore frameworks that extend beyond traditional human paradigm. Furthermore, by integrating concepts from astrobiology, investigators can develop frameworks that account for potential biological components of extraterrestrial artifacts, such as organisms or biological remnants that may have come from other worlds. This biological focus may also consider whether an extraterrestrial civilization relied upon biology-based technologies that we might not yet recognize or understand. Theoretical frameworks in extraterrestrial archaeology should acknowledge the limits of human knowledge and embrace diverse perspectives, as a truly comprehensive understanding of such artifacts demands openness to radical hypotheses and interpretations. By anchoring investigations in robust theoretical foundations, researchers can imbue their studies with intellectual rigor and foster innovative insights into the nature and implications of extraterrestrial relics.

Cultural and Symbolic Interpretations

Cultural and symbolic interpretations of extraterrestrial artifacts play a pivotal role in expanding our understanding of ancient civilizations and their interactions with potential alien visitors. The analysis of these relics not only sheds light on the technological advancements of past societies but also provides valuable insights into their belief systems, societal structures, and cultural practices. Understanding the cultural and symbolic significance of these artifacts is crucial in deciphering the narratives and mythologies associated with extraterrestrial contact and influence.

One approach to interpreting these artifacts involves cross-referencing them with known terrestrial cultural symbols and archaeological patterns. By identifying similarities and discrepancies, researchers can discern potential shared themes and iconographies between human cultures and any hypothetical extraterrestrial civilizations. This comparative analysis helps unravel the potential impact of extraterrestrial interaction on the development of ancient cultures and belief systems, offering tantalizing glimpses into the possible transcendence of cosmic knowledge between disparate worlds.

Furthermore, the study of cultural and symbolic interpretations extends beyond the tangible artifacts themselves. It encompasses the investigation of related artwork, written records, oral traditions, and architectural designs. By thoroughly examining the broader cultural context surrounding these artifacts, researchers can elucidate the significance of extraterrestrial influences on the creative expressions and sociocultural dynamics of ancient societies.

It is vital to recognize that interpretations of extraterrestrial artifacts should be approached with deep cultural sensitivity and cross-disciplinary collaboration. Cultural interpretations must involve input from anthropologists, historians, ethnographers, and experts in ancient languages and mythologies. Such collaborations ensure a comprehensive and nuanced understanding of the artifacts' cultural implications, allowing for the respectful integration of extraterrestrial narratives within the broader tapestry of human history and culture.

As we delve deeper into the cultural and symbolic interpretations of extraterrestrial artifacts, it becomes increasingly evident that these relics hold profound implications for our perception of humanity's place in the cosmos. By studying the cultural footprints left by potential extraterrestrial visitors, we embark on a thought-provoking journey that challenges conventional paradigms and invites us to contemplate the vastness of our interconnected cosmic heritage.

Summary and Conclusions on Extraterrestrial Artifact Studies

The study of extraterrestrial artifacts holds profound implications for our understanding of cosmic history and the potential interactions between advanced civilizations. Throughout this exploration, we have delved into the complexities of identifying, analyzing, and interpreting such relics, in an effort to expand our comprehension of the cosmos.

Highlighted within this discourse are the multifaceted approaches employed in discerning the enigmatic nature of extraterrestrial artifacts. By amalgamating various scientific disciplines, including archaeology, materials science, and anthropology, researchers have endeavored to unravel the mysteries encapsulated within these relics.

One of the primary resolutions to emerge from these studies is the necessity for a paradigm shift in understanding terrestrial history. The recognition that certain artifacts may possess non-terrestrial origins prompts critical reflection on human history and cultural development, humbling our place in the vast expanse of the cosmos.

Moreover, the implications extend beyond the mere identification and analysis of extraterrestrial artifacts. They instigate contemplations on the technological prowess of advanced alien civilizations, potentially offering unprecedented insights into their technological advancements and ingenuity. Furthermore, the temporal disparities unveiled through dating techniques not only challenge conventional chronologies but also prompt a reevaluation of humanity's place within the cosmic timeline.

The geographical distribution patterns further implicate the potential presence and influence of extraterrestrial beings on Earth, hinting at intricate connections that transcend our current understanding of planetary relations.

In essence, the culmination of these interpretations underscores the necessity for a holistic approach that integrates scientific methodologies, cultural insights, and philosophical considerations. By weaving these diverse strands of knowledge together, this multidisciplinary approach serves as a foundational framework for advancing the field of extraterrestrial artifact studies.

Ultimately, the pursuit of deciphering extraterrestrial relics not only enriches our comprehension of the universe but also challenges us to confront profound existential questions. Such inquiries illuminate our place in the cosmos and beckon us to shed new light on the enduring enigma of extraterrestrial influence. Thus, the journey through the nuances of extraterrestrial artifact studies presents an ongoing odyssey, propelling us toward new frontiers of intellectual inquiry and discovery.

Chapter Seven

Decoding Ancient Life: Tools and Techniques

Archaeogenetic Tools

The fusion of genetics and archaeology has revolutionized the study of ancient civilizations by providing unprecedented insights into the lives, migrations, and interactions of past populations. Archaeogenetic tools encompass a diverse array of techniques that enable researchers to delve into the genetic composition of ancient biological materials, shedding light on demographic patterns, kinship structures, and evolutionary dynamics. At the heart of this interdisciplinary field lies the extraction, amplification, and sequencing of ancient DNA (aDNA) from skeletal remains, mummified tissues, and archaeological artifacts. By unlocking the genetic information contained within these samples, scientists can unravel the genetic heritage of ancient individuals and populations, reconstruct their migratory pathways, and trace their connections to contemporary groups. Furthermore, the application of cutting-edge genomic technologies allows for the identification of genetic variations associated with specific traits, susceptibilities, and adaptations, offering clues about past lifestyles, diets, and disease susceptibilities. In addition to aDNA analysis, archaeogenetic tools encompass the study of ancient pathogens, parasites, and microbiomes, providing valuable insights into the health conditions, infections, and diseases prevalent in antiquity. This multidisciplinary approach not only illuminates the genetic diversity and biological affinities of ancient populations but also contributes to our understanding of the coevolutionary dynamics between humans and their microbial symbionts. The amalgamation of genetics and archaeology has the potential to address long-standing questions regarding cultural diffusion, population admixture, and technological innovations, offering a comprehensive view of the intricate tapestry of human history. In summary, the integration of archaeogenetic tools presents an unparalleled opportunity to explore the genetic legacies of our ancestors, uncover the complexities of their social structures, and decipher the impact of environmental factors on their genetic makeup.

Radioisotope Dating Techniques

Radioisotope dating techniques play a crucial role in unraveling the mysteries of ancient artifacts and archaeological sites. By utilizing the decay rates of radioactive isotopes, scientists can determine the age of materials with remarkable precision. One of the most widely used methods is radiocarbon dating, which relies on the decay of carbon-14 to estimate the age of organic remains, such as bone and charcoal, up to about 50,000 years ago. Beyond the scope of radiocarbon dating, other radioisotopes such as potassium-argon (K-Ar), uranium-lead (U-Pb), and rubidium-strontium (Rb-Sr) dating bring us even further back in time. These sophisticated techniques are essential for dating rocks, minerals, and fossils that extend far beyond the temporal limitations of radiocarbon dating. Each method has its unique strengths and limitations, making it imperative for archaeologists to carefully select the most appropriate technique for the material being analyzed. The process involves intricate sample preparation, precise measurement of isotopic ratios, and meticulous calculations to derive accurate age estimates. Additionally, the interdisciplinary nature of radioisotope dating involves collaboration between archaeologists, geologists, and physicists, highlighting the significance of cross-disciplinary approaches in scientific research. Constant refinements in analytical instruments and methodologies continue to enhance the reliability and accuracy of radioisotope dating techniques. This chapter aims to delve into the complexities and nuances of these indispensable tools, shedding light on their contribution to uncovering the temporal facets of our ancient world.

Molecular Clocks and DNA Analysis

The study of ancient DNA holds tremendous potential for unraveling the mysteries of human history, migration patterns, and evolutionary processes. Molecular clocks, a concept rooted in the observation that DNA accumulates mutations at a relatively constant rate, become essential tools for understanding the timeline of evolutionary events. By analyzing DNA sequences extracted from archaeological remains, researchers can trace the genetic changes that occurred over generations, providing insights into past populations and their relationships. This approach allows us to reconstruct ancestral lineages, estimate divergence times, and track the movement of ancient peoples across landscapes. Moreover, advancements in next-generation sequencing technologies have revolutionized the field of ancient DNA analysis, enabling scientists to retrieve genetic information from even poorly preserved samples. This has opened new avenues for studying prehistoric organisms, extinct species, and ancient pathogens, shedding light on the coevolution of humans and their environment. Furthermore, molecular genetic techniques can contribute to the identification of specific traits or adaptations that shaped ancient populations, offering a window into their physical attributes and biological affinities. The ethical dimensions of ancient DNA research also demand thoughtful consideration, particularly concerning the respectful treatment of human remains and the implications for Indigenous communities. As we delve deeper into the realm of molecular clocks and DNA analysis, it is crucial to uphold ethical standards and engage in meaningful dialogue with diverse stakeholders. The case studies presented in this chapter will showcase the practical applications of these techniques, illustrating their role in illuminating the threads of our shared history and enhancing our understanding of the complex tapestry of human existence.

Advanced Imaging Techniques in Archaeology

Advanced imaging techniques have revolutionized the field of archaeology by providing non-invasive methods to analyze and interpret ancient artifacts and sites. One of the key advancements in this area is the use of ground-penetrating radar (GPR), which allows researchers to detect buried structures, geological features, and artifacts without excavating the site. GPR works by emitting pulses of high-frequency electromagnetic waves into the ground and then recording the reflected signals, which can reveal subsurface anomalies and hidden archaeological remains. Another important imaging tool is LiDAR (Light Detection and Ranging), a remote sensing method that uses laser pulses to create detailed three-dimensional models of landscapes, uncovering previously unseen features such as ancient roads, structures, and landscapes that may be obscured by vegetation or human disturbances. In addition to GPR and LiDAR, thermal imaging has emerged as a valuable technique in archaeological investigations, enabling researchers to detect variations in temperature on the surface of an archaeological site, which can help identify subsurface structures and buried features based on differences in thermal conductivity. Moreover, advancements in multispectral imaging have allowed archaeologists to capture images beyond the visible spectrum, revealing details that are not discernible to the naked eye. This technique has been instrumental in identifying hidden paintings, inscriptions, and pigments on ancient artifacts. The integration of advanced imaging techniques with geographic information systems (GIS) has also enhanced the spatial analysis and visualization of archaeological data, enabling researchers to digitally map and analyze ancient landscapes and cultural heritage sites. Furthermore, the utilization of drones equipped with high-resolution cameras has facilitated aerial photography and photogrammetry, providing detailed topographic maps and 3D models of archaeological sites. These technological innovations have significantly expanded our ability to non-invasively study and preserve cultural heritage and have become indispensable tools in the modern archaeologist's toolkit.

Spectroscopy in Artifact Analysis

Spectroscopy, a powerful tool in the study of ancient artifacts, encompasses a wide range of techniques that enable us to analyze the composition and properties of materials. By employing principles of light-matter interactions, spectroscopic methods have revolutionized the field of archaeological analysis. Through the identification of elemental composition, molecular structure, and crystalline properties, spectroscopy provides invaluable insights into the origins, production techniques, and historical context of artifacts. One of the primary advantages of spectroscopic techniques lies in their non-destructive and non-invasive nature, allowing for the analysis of precious and delicate artifacts without causing damage. The utilization of various forms of spectroscopy, including Raman spectroscopy, X-ray fluorescence spectroscopy, and infrared spectroscopy, enables researchers to decipher the chemical composition of materials such as pigments, ceramics, metals, and glass. This, in turn, facilitates the determination of provenance, dating, authenticity, and technological advancements in ancient civilizations. Spectroscopic analyses not only aid in understanding the manufacturing processes and trade networks of ancient cultures but also provide critical information about artistic and decorative techniques. Furthermore, by identifying degradation mechanisms and alterations in artifacts over time, spectroscopy assists in conservation efforts, contributing to the preservation of our cultural heritage. The application of spectroscopic methods is not limited to terrestrial archaeology, as they can be adapted for extraterrestrial artifact analysis, offering potential insights into alien relics or traces of cosmic influences on Earth. With continued advancements in spectroscopic instrumentation and data analysis techniques, the role of spectroscopy in artifact analysis continues to expand, promising new avenues for unlocking the mysteries of the past.

Remote Sensing and Aerial Survey Methods

Remote sensing and aerial survey methods play integral roles in modern archaeological research, offering unprecedented opportunities to reveal hidden patterns and structures across the landscape. Through the use of cutting-edge technologies such as LiDAR (Light Detection and Ranging), multispectral imaging, and aerial photography, archaeologists can obtain comprehensive insights into ancient civilizations and their environmental impact. LiDAR, for instance, enables the creation of highly detailed digital elevation models, allowing researchers to virtually strip away vegetation and observe minute topographical features that might conceal buried archaeological sites. This non-invasive method has revolutionized the field by uncovering ancient settlements, road networks, and even ceremonial complexes that were previously unknown. Additionally, multispectral imaging provides invaluable data by capturing electromagnetic radiation beyond visible light, allowing the detection of subtle variations in soil composition or vegetation density that correspond to sub-surface cultural deposits. Aerial photography also remains indispensable for large-scale surveying, providing a bird's-eye view that aids in the identification of crop marks, soil discolorations, and other surface indications of past human activity. These advanced techniques not only enhance the efficiency of archaeological surveys but also contribute to a more comprehensive understanding of past landscapes and societies. Furthermore, they facilitate proactive conservation efforts and inform land-use planning by recognizing and protecting heritage sites threatened by modern development. In combination with ground-truthing and field verification, remote sensing, and aerial survey methods open new frontiers for investigating ancient civilizations, enabling researchers to piece together the complex tapestry of human history with greater accuracy and depth.

Underwater Exploration Technologies

Submerged landscapes hold a wealth of archaeological material, offering insights into ancient civilizations and environmental changes. Underwater exploration technologies play a pivotal role in uncovering and preserving these submerged historical treasures. Sonar imaging systems have revolutionized underwater archaeology by enabling researchers to map large areas with precision, identifying potential sites for exploration. Side-scan and multi-beam sonars provide high-resolution images of the seafloor, helping researchers locate and document submerged cultural remains such as shipwrecks, ancient ports, and settlements. Remotely Operated Vehicles (ROVs) equipped with specialized cameras and manipulator arms allow for detailed inspection and recovery of artifacts from depths that are challenging for human divers to access. Advanced diving technologies, such as closed-circuit rebreathers and mixed-gas diving, enable archaeologists to explore underwater sites at greater depths and for extended durations with reduced decompression risks. These tools bear significance in understanding past maritime activities, trade routes, and maritime connections between ancient civilizations. Additionally, advanced sediment coring systems and underwater excavation techniques help extract stratified deposits and organic remains from submerged contexts, shedding light on prehistoric environments and human behaviors. Furthermore, the use of Geographic Information Systems (GIS) and 3D visualization software aids in the reconstruction and interpretation of submerged landscapes, providing a comprehensive understanding of submerged archaeological sites and their contextual relationships. Challenges in underwater archaeological research include the conservation of recovered artifacts, as well as ethical considerations regarding site preservation and public access. Achieving a balance between exploration and preservation is vital in ensuring the responsible management of these invaluable resources. In summary, the integration of innovative underwater exploration technologies with traditional archaeological methods continues to expand the frontiers of knowledge, unveiling hidden chapters of human history that lie beneath the waves.

Data Management and Integrative Analytical Approaches

Effective data management and integrative analytical approaches are fundamental in advancing the field of cosmic archaeology. With the wealth of information generated from various technological tools and research methods, it is essential to implement robust data management strategies to ensure the integrity and accessibility of archaeological data. This section explores the intricacies of data management and highlights the significance of integrative analytical approaches in interpreting complex archaeological datasets.

Data management in cosmic archaeology involves the collection, storage, curation, and analysis of diverse datasets, including genetic, isotopic, geospatial, and cultural information. It encompasses the development of standardized protocols for data collection and metadata documentation to facilitate reproducibility and data sharing within the scientific community. Additionally, it underscores the importance of preserving archaeological data for future scholarly investigations and potential interdisciplinary collaborations.

Integrative analytical approaches play a critical role in synthesizing multidisciplinary data to extract meaningful insights about ancient life and extraterrestrial influences. By integrating genetic, geological, and cultural datasets, researchers can unravel intricate narratives about human evolution, environmental adaptations, and potential interactions with extraterrestrial phenomena. These approaches often involve cross-disciplinary collaboration among archaeologists, geneticists, geologists, and experts in remote sensing and imaging technologies to leverage diverse expertise in unraveling historical mysteries.

Furthermore, the integration of cutting-edge computational methods, such as machine learning algorithms and statistical modeling, enables researchers to discern patterns and correlations within vast archaeological datasets. This computational approach empowers scholars to conduct large-scale analyses and identify subtle relationships that may escape traditional qualitative assessments. Moreover, the integration of computational tools facilitates the visualization of complex data structures, aiding in the recognition of spatial and temporal patterns that underpin cosmic archaeological phenomena.

Ethical considerations also permeate data management and integrative analytical endeavors in cosmic archaeology. Maintaining ethical standards involves safeguarding the privacy and dignity of indigenous communities, respecting cultural heritage, and ensuring transparent practices in handling sensitive genetic and cultural information. Balancing scientific advancement with ethical responsibility is paramount in fostering mutual respect and collaboration with diverse stakeholders, including local communities, policymakers, and international organizations.

In summary, effective data management and integrative analytical approaches are pivotal components of advancing cosmic archaeology. Embracing rigorous data stewardship, interdisciplinary collaboration, and ethical guidelines fortifies the integrity and impact of archaeological research, ultimately enriching our understanding of ancient life and its cosmic context.

Ethical Considerations in Archaeological Research

Archaeological research is not merely a scientific endeavor, but also a field deeply intertwined with ethical considerations. As archaeologists uncover and analyze artifacts and remains from ancient civilizations, it is crucial to uphold ethical standards that respect the cultures and descendants tied to these discoveries. One primary ethical consideration is the principle of cultural sensitivity and respect for indigenous perspectives. In many cases, archaeological sites are located on lands that hold cultural significance for indigenous communities. Therefore, archaeologists need to engage with these communities, seek their input, and acknowledge their rights to heritage, especially when conducting research on the remains or artifacts of their ancestors. Furthermore, collaboration with indigenous groups can provide valuable insights and perspectives that enrich archaeological interpretations. Another critical ethical aspect involves the responsible stewardship of archaeological findings. This includes ensuring proper conservation, curation, and public accessibility of artifacts and archaeological sites. By preserving and making these findings accessible, researchers can contribute to educational and cultural enrichment while fostering a sense of shared heritage among diverse communities. In addition to upholding ethical responsibilities towards human remains and artifacts, archaeologists must also consider the impact of their work on the environment. Preservation of natural surroundings and ecosystems around archaeological sites is vital to maintaining the context of these historical places. Furthermore, ethical archaeology encourages sustainable practices that minimize environmental impact and promote conservation efforts. This ethical framework extends beyond the fieldwork itself to encompass the publication and dissemination of research findings. Ethical considerations dictate that researchers represent their findings accurately and respectfully, avoiding sensationalism and respecting the dignity of those whose legacies are being uncovered. Moreover, engaging with descendant communities and relevant stakeholders throughout the research process helps ensure that their perspectives are included and respected in the final interpretations and representations of the archaeological record. Overall, an ethical approach to archaeological research encompasses a commitment to cultural sensitivity, responsible stewardship, environmental conservation, and respectful engagement with stakeholders. By upholding these principles, archaeologists can conduct research that not only advances scientific knowledge but also fosters understanding, empathy, and appreciation for the rich tapestry of human history.

Applying Techniques in Fieldwork

The application of advanced archaeological techniques in fieldwork has significantly enhanced our understanding of ancient civilizations and their lifeways. This section highlights several intriguing case studies where these tools have been effectively employed to unravel the mysteries of the past.

In a remarkable excavation at a prehistoric site, archaeogenetic tools were instrumental in reconstructing the genetic history of an ancient population. By analyzing ancient DNA extracted from skeletal remains, researchers were able to trace the migration patterns and genetic diversity of the community, shedding light on their origins and interactions with neighboring groups. This groundbreaking discovery not only provided insight into the demographic history of the region but also presented a compelling narrative of human dispersal and cultural exchange.

Another captivating case study involved the use of sophisticated remote sensing technologies to uncover hidden archaeological features. Aerial survey methods, such as LiDAR imaging, revealed intricate details of ancient landscapes, including buried settlements, ceremonial structures, and agricultural systems. The high-resolution data obtained through these non-invasive techniques offered a comprehensive view of past human activities, revolutionizing our knowledge of ancient civilizations and their environmental adaptations.

Furthermore, a groundbreaking underwater exploration project utilized advanced sonar and imaging technologies to document submerged archaeological sites. By employing robotic submersibles and acoustic mapping, researchers mapped out a sunken city, providing unprecedented insights into maritime trade routes, seafaring cultures, and the impact of sea-level changes on coastal societies. These discoveries not only enriched our understanding of ancient maritime civilizations but also underscored the resilience of human communities in adapting to dynamic environments.

One of the most compelling applications of advanced imaging techniques in archaeology comes from the detailed analysis of intricate artifacts. Utilizing state-of-the-art techniques such as micro-CT scanning and 3D modeling, researchers were able to decode the manufacturing processes and functional properties of enigmatic objects. This in-depth examination revealed the technological sophistication of ancient craftsmen and offered new perspectives on the cultural significance and utilitarian purpose of these artifacts.

In summary, these captivating case studies demonstrate the transformative power of modern archaeological techniques in uncovering the rich tapestry of human history. By combining innovative tools with rigorous fieldwork, archaeologists continue to push the boundaries of knowledge, revealing extraordinary insights into the diverse and dynamic trajectories of ancient civilizations.

Chapter Eight

Exploring Alien Artifacts: Tracing Extraterrestrial Influence

The quest to understand and trace extraterrestrial influences through recovered artifacts is a compelling endeavor that transcends conventional archaeological inquiry. Fundamental to this pursuit is the exploration of foundational concepts aimed at defining the scope of extraterrestrial influence on human civilizations. This involves scrutinizing historical, archaeological, and scientific evidence with an interdisciplinary lens to grasp the profound implications of unearthed relics purportedly linked to alien origins. The investigation into extraterrestrial artifactual influence unfolds against the backdrop of humanity's enduring fascination with the possibility of contact with beings from beyond our planet. It calls for a meticulous analysis of alleged alien relics, which are imbued with enigmatic narratives that challenge orthodox perspectives on ancient civilizations and their technological capabilities. By delving into this uncharted terrain, we navigate intricate webs of speculation, skepticism, and curiosity, seeking to decipher the potential impact of extraterrestrial encounters on the evolution of human societies. At the crux of this inquiry lies the need to delineate the parameters of what constitutes an extraterrestrial artifact and how it intertwines with terrestrial archaeological records. This entails exploring the blurred boundaries between mundane objects of historical significance and those that hint at celestial origins, urging us to reevaluate established paradigms of human history and cultural development. As we embark on this intellectual odyssey, it becomes imperative to adopt a discerning approach that gracefully integrates traditional archaeological methodologies with cutting-edge scientific advancements. Esteemed scholars, maverick researchers, and inquisitive minds assemble to dissect the complexities of assessing extraterrestrial artifactual influence, thus propelling the discourse into realms once consigned to speculative conjecture. In contemplating these matters, one is compelled to ponder the profound ramifications of such revelations and their implications for humanity's collective understanding of its place in the cosmos. Ultimately, this section endeavors to lay a robust foundation for comprehending the intricacies and controversies surrounding the purported connection between humanity and extraterrestrial influences as manifested through recovered artifacts.

Historical Context: Discoveries of Alleged Alien Relics

Throughout history, numerous accounts and discoveries have fuelled speculation about the presence of alleged alien relics. Ancient civilizations across the globe have left behind artifacts that bear enigmatic designs, inexplicable materials, and seemingly advanced technological features—elements that some have attributed to extraterrestrial influence. In Mesopotamia, the Sumerians depicted their gods with a striking resemblance to modern depictions of aliens, sparking theories of ancient contact. Similarly, the mysterious Nazca Lines in Peru, believed to have been created between 500 BCE and 500 CE, have spurred conjecture about their potential link to extraterrestrial visitors due to their massive scale and representations of otherworldly figures. The iconic Easter Island statues, or moai, have also raised eyebrows due to their remarkable craftsmanship and remote location, evoking questions about potential extraterrestrial assistance. In more recent history, the discovery of artifacts like the Antikythera Mechanism—a complex ancient Greek analog computer—challenges conventional narratives of human technological evolution, prompting alternative hypotheses including extraterrestrial intervention. These instances represent just a fraction of the many historical puzzles that continue to captivate the imagination and inspire research into the possible connection between human history and unearthly influences. As we delve into these historical conundrums, it becomes increasingly apparent that the quest to uncover the truth behind alleged alien relics is ripe with intellectual intrigue and potential for groundbreaking discoveries. Through rigorous investigation and multidisciplinary collaboration, this chapter aims to navigate the labyrinth of the historical context surrounding alleged alien relics, shedding light on the enduring mysteries that have persisted throughout the annals of human civilization.

Archaeometric Techniques in Extraterrestrial Artifact Analysis

Archaeometric techniques play a pivotal role in unraveling the enigma surrounding extraterrestrial artifacts. These methodologies blend advanced scientific disciplines with traditional archaeological approaches to scrutinize and comprehend the material composition, age, and origins of these enigmatic objects. The utilization of cutting-edge tools such as scanning electron microscopy, X-ray diffraction, and spectrometry enables researchers to delve into the atomic and molecular structure of alleged alien relics. Isotopic analysis provides insights into the isotopic composition of materials, offering valuable clues about their terrestrial or non-terrestrial provenance. Furthermore, radiocarbon dating and luminescence dating techniques are employed to establish the age of unearthed extraterrestrial objects and assess their chronology within cosmic timescales. Additionally, thermoluminescence dating offers a method to ascertain the last exposure of certain crystalline materials to heat or sunlight, further enriching our understanding of these artifacts. Neutron activation analysis is another indispensable tool that facilitates the determination of trace elements, shedding light on the geographical sources of these objects. Moreover, the application of non-destructive testing methods like computed tomography (CT) scanning allows for comprehensive internal structural analysis of complex extraterrestrial artifacts without compromising their integrity. Raman spectroscopy aids in identifying the molecular and crystal structures of samples, providing essential information regarding their elemental composition and potential extraterrestrial origins. It is imperative to acknowledge that the interdisciplinary nature of archaeometric techniques fosters a holistic and rigorous approach to interrogating alleged alien relics. Through the convergence of diverse analytical methods, researchers can navigate the complexities inherent in investigating objects purported to have extraterrestrial provenance. This multidisciplinary integration not only elevates the scientific rigor of the analyses but also underscores the significance of collaborative endeavors in advancing our understanding of potential extraterrestrial influences on Earth's history.

Typology of Alien Artifacts

Alien artifacts, with their enigmatic origins and astonishing diversity, present a significant challenge in classification and typology. This section endeavors to establish a framework for categorizing these otherworldly objects, drawing upon comparative analysis, material attributes, and historical contexts. The typology of alien artifacts can be delineated based on various criteria, such as form, function, composition, and cultural significance. By discerning patterns and relationships among these extraterrestrial relics, it becomes feasible to construct a taxonomy that facilitates a deeper understanding of their potential origins and purposes. Furthermore, the typological approach offers valuable insights into the technological and sociocultural aspects of alien civilizations, shedding light on their cognitive processes and creative expressions. To commence the typological analysis, it is imperative to scrutinize the different forms and structures manifested in alien artifacts. This entails cataloging a diverse array of objects, ranging from intricately designed tools and utensils to intricately decorated ornaments and symbolic representations. Each category unveils distinctive characteristics that offer crucial clues about the technological prowess and aesthetic sensibilities of the extraterrestrial societies that engendered them. Concurrently, the functional analysis of alien artifacts elucidates their purported uses and practical applications, corroborating the conjectures drawn from their physical attributes. Additionally, the compositional analysis delves into the elemental constituents and isotopic ratios of these artifacts, seeking disparities from terrestrial materials that might betray their non-Earthly origins. Such meticulous examinations serve as a cornerstone for establishing a robust typology rooted in scientific rigor. Moreover, by integrating the historical provenance of these artifacts into the typological framework, it becomes viable to trace the evolution and diffusion of extraterrestrial influences across different temporal and spatial milieus. This multifaceted typological approach not only enriches our comprehension of alien artifacts but also paves the way for meaningful comparisons with terrestrial material culture, affording deeper insights into the cross-cultural interactions between humanity and hypothetical extraterrestrial entities.

Material Science Insights: Composition and Non-Earthly Origins

Material science offers a crucial lens for scrutinizing the nature of extraterrestrial artifacts, presenting an opportunity to delve into their composition and origins. Through advanced spectroscopic and microscopic analyses, researchers can unravel the elemental makeup of these enigmatic objects, shedding light on their non-earthly origins. By comparing the material properties of alien artifacts with those found in known terrestrial materials, scientists can discern distinct signatures indicative of extraterrestrial provenance. This exploration prompts questions regarding the mechanisms that govern the formation and preservation of these anomalous materials, transcending conventional paradigms of archaeological investigation.

Furthermore, the interdisciplinary amalgamation of material science and archaeology enables the identification of anomalous isotopic ratios and exotic atomic configurations within these artifacts. Such revelations challenge established paradigms of material formation and provide compelling evidence for non-terrestrial origins. The integration of cutting-edge mass spectrometry, X-ray diffraction, and electron microscopy proves instrumental in deconstructing the inherent properties of these alien relics, fostering a comprehensive understanding of their unique material qualities and divergent genesis.

Moreover, the quest to decipher non-earthly origins necessitates a meticulous evaluation of isotopic anomalies and crystallographic structures prevalent in extraterrestrial artifacts. These distinctive characteristics confound traditional geological and mineralogical frameworks, propelling researchers towards a paradigm shift in conceptualizing the process and conditions under which these unconventional materials crystallized. As such, material science not only substantiates the extraterrestrial origins of these artifacts but also introduces novel avenues for redefining the fundamental tenets of material synthesis and preservation beyond terrestrial bounds.

In essence, the application of material science methodologies in the analysis of extraterrestrial artifacts transcends the boundaries of conventional archaeological inquiry, leading to profound insights into their composition and non-earthly origins. This multidisciplinary approach offers an unparalleled vantage point for unraveling the perplexing enigma of ancient extraterrestrial relics, heralding a new epoch of scientific discovery and redefining humanity's perception of its place within the cosmos.

Cultural Comparisons: Style and Iconography in Alien Objects

Understanding the cultural context of alien artifacts is a complex and challenging endeavor. While the study of extraterrestrial objects poses unique obstacles due to their non-terrestrial origins, it also offers a fascinating window into the potential diversity of civilizations beyond Earth. Cultural comparison of style and iconography in these alien artifacts involves a nuanced examination of forms, motifs, and symbols that may hold significant cultural and historical meanings. Archaeologists and astronomers must collaborate to decipher the visual language embedded in these objects and draw parallels with known human cultures. By analyzing the artistic and symbolic elements present in alien artifacts, researchers can attempt to decipher the aesthetic and conceptual frameworks of extraterrestrial civilizations.

Comparative analysis will involve studying the similarities and divergences between the iconography found in alien artifacts and that of various terrestrial cultures throughout history. This process requires a meticulous examination of design elements, iconographic motifs, and stylistic features to identify potential correspondences and distinctions. Furthermore, the investigation will explore the potential universality of certain symbols or artistic conventions observed in extraterrestrial objects, shedding light on the possible existence of cross-cultural influences or shared mythological narratives among diverse cosmic societies.

Moreover, the study of style and iconography in alien objects necessitates an understanding of the technological, social, and environmental contexts within which these artifacts were produced. By examining the material culture and the broader societal dynamics reflected in these artifacts, researchers can gain insights into the technical expertise, religious beliefs, and sociopolitical structures of extraterrestrial civilizations. This multifaceted approach will enable a comprehensive interpretation of the cultural significance and functionality of alien artistic expressions, enriching our understanding of extraterrestrial societies and their interactions with the cosmos.

In summary, the comparative analysis of style and iconography in alien objects represents a pivotal contribution to the interdisciplinary field of cosmic archaeology. It not only facilitates the reconstruction of extraterrestrial cultural landscapes but also provides invaluable opportunities to engage in cross-cultural dialogues and broaden our perspectives on the potential diversity of intelligent life beyond our planet.

Interdisciplinary Approaches: Merging Archaeology and Astronomy

The merging of archaeology and astronomy represents a formidable interdisciplinary approach that has garnered increasing attention within the scientific community. This convergence is underpinned by the recognition that celestial phenomena have significantly impacted human societies throughout history, shaping cultural practices, belief systems, and monumental constructions. By integrating astronomical knowledge and archaeological methods, researchers seek to unravel the profound interconnectedness between the cosmos and terrestrial civilizations. Drawing from diverse disciplines, this integrative framework offers a multifaceted lens through which to interpret ancient artifacts and architectural structures, shedding light on their cosmological significance. The cross-pollination of astronomy and archaeology enables the identification of celestial alignments encoded within ancient sites, such as megalithic complexes and pyramids, indicative of a sophisticated understanding of astronomical principles among ancient cultures. Furthermore, the study of ancient mythologies and rituals in conjunction with celestial events provides invaluable insights into how prehistoric communities perceived and interacted with the cosmos. Methodologies encompassing landscape archaeology, spatial analysis, and remote sensing are employed to explore the temporal and spatial dimensions of astronomical alignments, offering compelling evidence of deliberate cosmic orientations. Key to this interdisciplinary approach is the utilization of advanced technologies, including LiDAR (Light Detection and Ranging) and satellite imagery, which enhance the precision and scope of archaeological investigations, uncovering hidden celestial connections embedded in the world's heritage. Moreover, through collaborative initiatives, such as astro-archaeoastronomy projects, experts synergize their expertise to advance the understanding of ancient cosmic knowledge, its transmission, and its enduring impact on human societies. The implications of this interdisciplinary endeavor extend beyond academic realms, enriching our appreciation of the profound intellectual and spiritual engagements of past civilizations with the celestial realm. Ultimately, the merger of astronomy and archaeology facilitates a holistic comprehension of the human-cosmos relationship, elucidating the enduring influence of astronomical phenomena on the cultural and technological evolution of humanity.

Theoretical Frameworks: Understanding Extraterrestrial Impact

The search for extraterrestrial artifacts necessitates a robust theoretical framework to comprehend the potential impact of alien influences on human history and culture. This chapter delves into the multifaceted theoretical underpinnings that guide our understanding of extraterrestrial artifacts and their implications. At its core, the theoretical framework encapsulates various perspectives from archaeology, astronomy, astrobiology, and cultural studies, amalgamating these diverse disciplines to construct a comprehensive lens through which to view the enigmatic remnants of potential extraterrestrial civilizations. The overarching objective is to elucidate the possible narratives and contexts within which alien artifacts could have shaped human development. One paradigm centers on the idea of panspermia, positing that life—and by extension, cultural dissemination—could have been facilitated through interstellar carriers such as cometary or meteoritic impacts. Alternatively, the principles of directed panspermia entertain the notion that deliberate transmission of microbial life or cultural information from an advanced extraterrestrial civilization may have occurred, thereby exerting a direct influence on terrestrial evolution. Moreover, the concept of cosmic ancestry explores the hypothetical genetic and cultural connection between Earth and celestial bodies, fostering the exploration of potentially shared origins. Theoretical frameworks are also influenced by considerations of ancient astronaut theories and the prospect of direct extraterrestrial visitation. These models invoke contentious yet thought-provoking discussions regarding potential alien interaction with ancient societies, thus shaping sociocultural narratives and belief systems. Additionally, the examination of psychological, sociological, and anthropological dimensions is imperative when attempting to understand how the awareness of extraterrestrial influence impacts human societies. It is crucial to explore the potential ramifications on belief systems, identity formation, and societal structures. Understanding how these theoretical paradigms intersect and diverge provides a nuanced grasp of how various domains of knowledge coalesce to enrich our comprehension of potential extraterrestrial artifacts and their societal significance. By embracing these conceptual frameworks, we can attentively navigate the complexities of interpreting extraterrestrial influence and its profound implications for humanity's collective narrative.

Case Study Review: Significant Findings and Their Implications

This section unearths a comprehensive overview of notable case studies that have significantly contributed to the understanding of alien artifact exploration. Each case study is meticulously examined, providing insights into the profound implications of their findings on our understanding of extraterrestrial influence. The Kecksburg Incident, an alleged UFO crash in Pennsylvania in 1965, remains one of the most iconic cases in UFO lore. Its investigation not only shed light on the societal response to potential extraterrestrial contact but also prompted discussions on government secrecy and misinformation. Similarly, the Roswell Incident continues to captivate public intrigue, reflecting the enduring fascination with alleged extraterrestrial encounters. The examination of purported debris from a crashed UFO in Roswell, New Mexico, has ignited debate and conspiracy theories, influencing popular culture and scientific inquiry alike. Moving beyond terrestrial confines, the Apollo 11 mission's enigmatic encounter with anomalous structures on the moon has spurred speculation about ancient extraterrestrial artifacts. By critically evaluating these pivotal case studies and their enduring impact, this section illuminates the complexities and controversies surrounding the investigation of alleged extraterrestrial artifacts. Furthermore, it underscores the importance of nuanced analysis and interdisciplinary collaboration in approaching such extraordinary claims. These case studies serve as cautionary tales while simultaneously inspiring further research into the enigmatic realm of alien artifacts, encouraging scholarly rigor and open-minded inquiry into the mysteries of cosmic archaeology.

Summary and Conclusions: The State of the Field and Future Avenues

As we conclude our exploration of alien artifacts and their implications, it becomes clear that this field is at a critical juncture. The significant findings from various case studies have shed light on the potential existence of extraterrestrial influence on our planet. However, the state of the field also reflects the dearth of conclusive evidence in this domain. It is imperative to approach this subject with cautious skepticism while remaining open to the possibility of groundbreaking discoveries.

Moving forward, the future avenues for research in alien artifact exploration appear promising yet challenging. Advancements in technology, particularly in spectroscopy, isotopic analysis, and material science, offer unprecedented opportunities to delve deeper into the composition and origins of alleged extraterrestrial relics. Furthermore, interdisciplinary collaboration between archaeologists, astronomers, astrophysicists, and astrobiologists will be essential to gain comprehensive insights into the potential interstellar influences on Earth's cultural and technological development.

The state of the field demands a balanced approach that amalgamates scientific rigor with an open-minded exploration of unconventional theories and hypotheses. In doing so, the field of exploring alien artifacts can transcend its current limitations and contribute to a broader understanding of our place in the cosmos.

Looking ahead, the evolution of space exploration and ongoing endeavors such as the Search for Extraterrestrial Intelligence (SETI) present tantalizing prospects for uncovering artifacts or traces of extraterrestrial civilizations. These efforts not only hold scientific significance but also pose profound philosophical and existential questions about humanity's position in the universe.

In conclusion, the quest to trace extraterrestrial influence through the exploration of alien artifacts stands at the intersection of scientific inquiry and speculative exploration. The state of the field beckons researchers to embrace a sense of curiosity tempered with critical reasoning as they embark on this intellectual odyssey. The future avenues for research in this domain hold the promise of reshaping our understanding of terrestrial and extraterrestrial interactions, offering a glimpse into the profound mysteries of the cosmos.

Chapter Nine

Cosmic Anomalies and Terrestrial Parallels

The field of cosmic anomalies offers a fascinating array of mysterious phenomena that challenge our current understanding of the universe. These anomalies range from unexplained changes in celestial radiation to unusual patterns in the distribution of matter across galaxies. Such unexplained events, observed in the vastness of space, cover a broad spectrum of cosmic oddities. The variety and extent of these anomalies have significant implications for how we understand the universe, pushing scientists to explore deeper into the complexities of astrophysics and cosmology.

Within the cosmic domain, anomalies manifest as deviations from expected patterns or behaviors established by prevailing scientific models. These anomalies elude conventional explanations and often prompt scientists to reassess fundamental assumptions regarding the functioning of cosmic structures and processes. They encompass a diverse array of perplexing events, including unexpected emissions of high-energy particles, bizarre gravitational lensing effects, and peculiar disturbances in cosmic microwave background radiation.

Emerging from both conceptual and observational perspectives, the study of cosmic anomalies demands an interdisciplinary approach that seamlessly integrates expertise from fields such as astronomy, theoretical physics, and data science. This collaborative effort is essential for comprehensively unraveling the underlying mechanisms driving these enigmatic phenomena and discerning their potential influence on the broader cosmic framework.

The exploration of cosmic anomalies is not merely a pursuit of esoteric curiosities; rather, it underscores the intricate interplay between known physical laws and unforeseen cosmic occurrences. By critically examining these anomalies, we gain insights into the frontiers of scientific knowledge and are compelled to explore new theoretical avenues that can accommodate these outliers within the grand narrative of cosmic evolution. Our efforts to understand cosmic anomalies also pose vital questions about the limits of human perception and comprehension, prompting introspection into the very nature of scientific inquiry and the boundaries of empirical validation.

As we embark on a journey through the manifold manifestations of cosmic anomalies, we are called to navigate through the uncharted territories of the universe, where each anomaly serves as a signpost guiding us toward deeper revelations about the fabric of reality itself.

Defining Anomalies in a Cosmic Context

Anomalies within the cosmic framework are often enigmatic phenomena that defy conventional explanations within the existing paradigms of astrophysics and cosmology. These occurrences present intriguing puzzles that challenge our understanding of the universe. Unlike regular astronomical events, cosmic anomalies may manifest as unanticipated celestial observations, inexplicable energy signatures, or perplexing spatial structures that elude conventional theoretical frameworks.

In the quest to comprehend these elusive anomalies, it becomes imperative to delineate their characteristics within the cosmic context. Often, anomalies are characterized by their deviation from expected patterns, exhibiting behaviors that deviate significantly from established astrophysical norms. These deviations may manifest as unusual luminosity variations, irregular gravitational lensing effects, unexplained cosmic ray patterns, or peculiar astronomical spectral signatures that warrant meticulous investigation.

Furthermore, anomalies extend beyond mere observational discrepancies, encompassing phenomena that challenge fundamental physical laws or postulated cosmic processes. These can include instances of apparent superluminal motion, unexpected periodic fluctuations in cosmic phenomena, or unexplained cosmic-scale synchronicities that transcend prevailing interpretations of space-time dynamics. The multifaceted nature of cosmic anomalies necessitates a comprehensive approach to define and classify these perplexing occurrences systematically.

The classification of anomalies in cosmic contexts involves discerning between transient observational irregularities and enduring enigmas that persist in defiance of successive scientific inquiries. By categorizing anomalies based on their temporal persistence, emergent patterns, and causal mechanisms, researchers endeavor to unravel the underlying cosmic complexities. Such an approach facilitates the identification of recurring cosmic irregularities that exhibit stable behavioral attributes, thereby providing critical insights into potentially groundbreaking cosmic phenomena.

Moreover, defining anomalies in cosmic contexts demands careful consideration of interdisciplinary perspectives, drawing upon insights from diverse scientific domains such as theoretical physics, quantum mechanics, and astronomical instrumentation. This integrative approach seeks to encapsulate the multifaceted dimensions of anomalies, transcending conventional disciplinary boundaries to foster a holistic understanding of these enigmatic cosmic manifestations.

In essence, the process of delineating anomalies in a cosmic context encompasses an intricate tapestry of observational, theoretical, and methodological considerations. As we strive to comprehend these cosmic conundrums, engaging in a rigorous analysis of anomalies holds the promise of unveiling hitherto unexplored facets of the cosmos, propelling us towards new frontiers of knowledge and discovery.

Unexplained Cosmic Events

In the study of cosmic anomalies, delving into case studies of unexplained cosmic events offers invaluable insights into the mysterious occurrences that defy conventional scientific explanations. These case studies serve as windows into the enigmatic facets of our universe, prompting deep contemplation and rigorous investigation. Within the realm of cosmic archaeology, the exploration of unexplained phenomena is particularly compelling, as it unveils tantalizing clues about the cosmic tapestry. One such case study involves the puzzling phenomenon of fast radio bursts (FRBs), transient radio pulses of extragalactic origin discovered through radio astronomy. The sporadic nature and extraordinary energy levels of FRBs challenge existing astrophysical models, sparking fervent debate among researchers. Another captivating case study centers on the enigmatic Tabby's Star, also known as Boyajian's Star, which displays irregular and unexplained dimming patterns. Astrophysicists have proposed various hypotheses, ranging from alien megastructures to dust clouds, but the true nature of this celestial enigma remains elusive. Moreover, the anomalous acceleration of the Pioneer spacecraft has confounded scientists, defying the predictions of classical physics and prompting renewed scrutiny of gravitational theories. In each of these cases, the intricate interplay between observation and theory underscores the complex nature of cosmic anomalies, compelling us to reassess our understanding of the universe. As researchers meticulously dissect these cosmic enigmas, they find terrestrial parallels in historical accounts of inexplicable events, emphasizing the enduring human fascination with the unknown. It is through these case studies that we are impelled to navigate the frontiers of cosmic knowledge, fostering a deeper appreciation for the enigmatic phenomena that pervade the cosmos.

Terrestrial Parallels: Echoes on Earth

The exploration of cosmic anomalies often draws intriguing parallels with terrestrial events and phenomena, prompting a deeper examination of potential connections. Terrestrial parallels offer valuable insights into understanding cosmic anomalies by contextualizing them within the framework of familiar occurrences on Earth. By scrutinizing historical, geological, and meteorological data, researchers have uncovered compelling correlations between unexplained cosmic events and analogous incidents in our planet's history.

One notable example lies in the investigation of meteorite impacts and their terrestrial parallels. The study of impact craters has revealed striking resemblances to certain geological formations on Earth, shedding light on the potential effects of extraterrestrial impacts. By examining the geological records left by meteorite strikes, scientists can discern similarities in the aftermath of cosmic collisions and catastrophic terrestrial events.

Moreover, investigations into electromagnetic disturbances and cosmic radiation have spurred comparative analyses with terrestrial occurrences, such as geomagnetic storms and atmospheric anomalies. By delving into these parallels, researchers aim to discern common patterns and elucidate potential causal links between cosmic and terrestrial phenomena.

Another crucial avenue of exploration involves the analysis of ancient texts, myths, and folklore that recount events resembling cosmic anomalies.

Through interdisciplinary studies intersecting archaeology, anthropology, and cosmology, researchers seek to unravel the cultural reverberations of such phenomena and their enduring impact on human societies.

Further parallels emerge from the study of extremophilic organisms, which thrive in environments reminiscent of planetary conditions mirroring those found in space. By studying these earthly analogs, researchers gain crucial insights into the potential habitability of extraterrestrial environments and the survival mechanisms of life forms in extreme cosmic settings.

In essence, the exploration of terrestrial parallels offers a multifaceted approach to comprehending cosmic anomalies, allowing for cross-disciplinary investigations that draw upon diverse fields of expertise. By bridging the realms of cosmic and terrestrial phenomena, researchers endeavor to uncover underlying connections that may illuminate the enigmatic tapestry of cosmic anomalies and expand the horizons of scientific inquiry.

Methodological Approaches for Studying Anomalies

In the pursuit of unlocking the mysteries behind cosmic anomalies and their terrestrial parallels, it becomes imperative to carefully delineate methodological approaches that can underpin a rigorous investigation. The study of anomalies demands an interdisciplinary framework that amalgamates the tools and techniques of various scientific domains. At the crux of this endeavor lies the need for a holistic and systematic methodological approach. Establishing a standardized methodology for anomaly study involves meticulous planning and implementation, integrating methodologies from fields such as astrophysics, cosmology, archaeoastronomy, geophysics, and other allied disciplines. This cross-disciplinary synergy serves as the cornerstone in elucidating anomalous phenomena, fostering collaboration between diverse experts with multifaceted perspectives. The holistic analysis entails the fusion of qualitative and quantitative methodologies, each playing a pivotal role in comprehending the nature and implications of anomalies. Qualitative methods like ethnographic studies, historical analysis, and phenomenological inquiries aid in contextualizing anomalies within cultural, historical, and societal realms. These approaches illuminate the human experiences and interpretations of cosmic events, enriching the understanding of their impact on civilizations across time. In contrast, quantitative methods encompass statistical analyses, computational modeling, geospatial mapping, and empirical observations, offering tangible data-driven insights into the spatial, temporal, and causal dimensions of anomalies both in cosmic and terrestrial contexts. Exemplary techniques include the application of advanced imaging technologies, geophysical surveys, remote sensing, and spectroscopic analyses to scrutinize material artifacts or cosmic occurrences. Additionally, the utilization of cutting-edge instruments and telescopic observations enables astronomers and astrophysicists to capture enigmatic stellar phenomena, thereby unraveling celestial anomalies. Sequentially, a vital attribute in anomaly investigation is the critical examination of anomalous phenomena through controlled experimentation, wherein hypothetical models are rigorously tested against empirical observations and predictive scenarios. Envisioned hypothetical models invariably serve as a springboard for the simulation of anomalous cosmic events, allowing researchers to deduce potential underlying mechanisms and unveil enigmatic patterns. Furthermore, adopting a comparative approach to analyze terrestrial analogs of cosmic anomalies facilitates the discernment of recurring patterns and universal principles governing anomalous phenomena. It necessitates scrutinizing geological, archaeological, and climatological records to discern congruences between purported cosmic anomalies and their terrestrial counterparts. This comparative analysis orchestrates a web of interconnections between disparate disciplines, unearthing intriguing revelations about the intrinsic interplay of cosmic and terrestrial systems. Ultimately, a robust methodological repertoire serves as the linchpin for transcending the nebulous borders of anomalies and forging systematic pathways to comprehend the enigmatic tapestries of the cosmos.

Cross-Disciplinary Perspectives

In the exploration of cosmic anomalies and their terrestrial parallels, it becomes imperative to harness the insights from a myriad of disciplines. Cross-disciplinary perspectives offer a unique vantage point, allowing researchers to leverage diverse methodologies, theories, and empirical evidence to unravel the enigmatic connections between cosmic and terrestrial phenomena. From astrophysics to archaeology, from geology to astrobiology, each discipline brings its own set of tools and frameworks that enrich the investigation of these anomalies.

Astrophysical perspectives illuminate the celestial origins of anomalous events, probing the fundamental forces and interactions that govern our universe. By peering into the depths of space and time, astrophysicists strive to discern the underlying mechanisms behind unexplained cosmic occurrences, shedding light on the intricate web of cosmic phenomena that shape our cosmos. Meanwhile, archaeological insights offer a lens into the deep-time history of Earth, unveiling traces of past events or influences that resonate with cosmic anomalies, and presenting intriguing intersections between human history and cosmic phenomena.

Geological viewpoints provide a tangible link between cosmic events and terrestrial manifestations, offering clues embedded within the Earth's geological record that mirror or intersect with cosmic anomalies. The study of geological formations, isotopic compositions, and sedimentary layers presents an opportunity to identify terrestrial proxies of cosmic events, fostering a holistic understanding of the interconnectedness between the Earth and the cosmos. Furthermore, astrobiological perspectives inject a dimension of life and habitability, exploring the potential implications of cosmic anomalies on the search for extraterrestrial life and the conditions that foster it in the universe.

Moreover, the integration of interdisciplinary frameworks spurs innovation in explicating the underlying causes and correlates of cosmic anomalies, fostering collaborations that transcend traditional academic boundaries. Interdisciplinary dialogues cultivate a fertile ground for synthesizing a comprehensive narrative that harmonizes disparate strands of evidence and conjectures, propelling the quest to decipher cosmic mysteries forward. This convergence of disciplines not only broadens the intellectual scope but also engenders a heightened awareness of the interconnectedness of cosmic and terrestrial realms, illuminating the profound implications that these anomalies pose for the fabric of our understanding of the universe.

The cross-pollination of ideas, methodologies, and findings from varied disciplines contributes to a rich tapestry of knowledge and insight, reinforcing the notion that the study of cosmic anomalies necessitates an inclusive, multidimensional approach. As we navigate this interdisciplinary terrain, examining the entangled skein of cosmic and terrestrial phenomena, we kindle fresh perspectives that have the potential to revolutionize our comprehension of the universe and our place within it.

Theoretical Models Linking Terrestrial and Cosmic Phenomena

In the exploration of cosmic anomalies and their terrestrial parallels, the development of theoretical models is crucial for establishing a framework to understand the interconnected nature of these phenomena. Theoretical models serve as the conceptual scaffolding that allows researchers to organize observations and experimental data into coherent explanations. When it comes to linking terrestrial and cosmic phenomena, the challenge lies in identifying common threads that can bridge the scales of our planet to the vastness of the cosmos. One such model involves the concept of fractal geometry applied to cosmological structures and earthly patterns. Fractals, with their self-similar shapes repeating at different scales, offer a way to visualize how cosmic phenomena may find expression in seemingly unrelated terrestrial formations. This approach has yielded insightful parallels in phenomena such as the branching patterns of galaxies mirroring the structure of trees or river networks on Earth. Another model draws from complexity theory, suggesting that certain cosmic processes exhibit dynamics akin to those found in complex systems on Earth. By studying emergent behaviors and self-organization across various scales, researchers have uncovered intriguing parallels between phenomena like solar flares and nerve cell firing. Importantly, these theoretical models not only facilitate the identification of parallels but also inform predictive frameworks that could lead to new discoveries. Additionally, the burgeoning field of quantum biology presents a promising avenue for understanding how quantum phenomena at the cosmic scale may manifest in biological systems on Earth. The application of quantum principles to biological processes raises profound questions about the relationship between life and the cosmos, potentially shedding light on how fundamental aspects of existence are linked across scales. Furthermore, theoretical models that explore the influence of celestial events on terrestrial phenomena, such as climate shifts or evolutionary patterns, can elucidate the broader implications of cosmic connections. By integrating insights from diverse fields, including astrophysics, geology, biology, and anthropology, these models aim to construct a comprehensive framework that transcends disciplinary boundaries. Through the synthesis of theoretical and observational approaches, researchers are poised to uncover new layers of unity within the cosmic tapestry, redefining our understanding of the interconnectedness between the terrestrial and the cosmic.

Implications for Current Scientific Understandings

The exploration of cosmic anomalies and their parallels with terrestrial phenomena presents profound implications for our current scientific understanding. As researchers endeavor to bridge the gap between cosmic and terrestrial events, it becomes increasingly evident that existing scientific frameworks may require reevaluation and expansion. The identification of correlations between unexplained cosmic occurrences and analogous terrestrial manifestations challenges traditional disciplinary boundaries and beckons a multidisciplinary approach. This necessitates a paradigm shift in scientific thinking, compelling scholars to adopt an integrative perspective that transcends entrenched silos of knowledge.

By scrutinizing the interconnectedness of cosmic and terrestrial phenomena, scientists are poised to not only deepen their comprehension of the universe but also redefine established theories. This convergence of insights from cosmology and archaeology stands to refine prevalent scientific paradigms and propel holistic inquiries into the nature of reality itself. Moreover, as anomalies revealed by cosmic investigations find uncanny mirrors on Earth, the foundational concepts of causality, determinism, and predictability in natural processes demand reconsideration. The newfound interplay between cosmic anomalies and terrestrial parallels augurs the necessity for a comprehensive revision of scientific constructs to accommodate these profound revelations.

Concurrently, the implications for current scientific understandings extend beyond theoretical reconceptualizations. The tangible impact encompasses interdisciplinary collaborations, as experts from diverse fields converge to unlock the mysteries encapsulated within cosmic anomalies and terrestrial echoes. This synergy not only enriches scientific discourse but also cultivates a symbiotic exchange of methodologies, thereby enriching the investigative arsenal available to both cosmologists and archaeologists. Additionally, the integration of approaches fosters intellectual cross-fertilization, giving rise to innovative research trajectories and pioneering analytical tools capable of delineating hitherto inscrutable domains of inquiry.

Furthermore, the revelation of commonalities between cosmic anomalies and terrestrial benchmarks engenders a paradigmatic evolution in our understanding of natural systems. The underlying unity governing seemingly disparate phenomena underscores the need for a conceptual metamorphosis in scientific epistemology. It impels scientists to transcend reductionist perspectives and embrace a holistic worldview that acknowledges the intrinsic connectedness of all cosmic and terrestrial processes, heralding a new era in human cognition and understanding.

Challenges in Correlating Cosmic and Terrestrial Data

Correlating cosmic and terrestrial data presents a multitude of complex challenges that require interdisciplinary collaboration and innovative methodologies. One of the foremost obstacles is the inherent disparity in scales between cosmic phenomena and terrestrial observations. While cosmic anomalies may manifest on astronomical scales, detecting and correlating these events with terrestrial data often involves reconciling vastly different measurement scales and units. Additionally, the chaotic and unpredictable nature of cosmic events poses challenges in identifying consistent patterns or correlations with terrestrial data. This necessitates the development of robust statistical methods and analytics to discern meaningful relationships amidst inherent variability. Furthermore, the interpretation of cosmic data in conjunction with terrestrial evidence demands a careful consideration of temporal and spatial factors. The timeframes associated with cosmic phenomena may span millions or even billions of years, necessitating creative approaches to align such vast timescales with the relatively succinct historical records available for terrestrial investigations. Similarly, spatial dimensions pose significant challenges, as celestial distances and stellar dynamics introduce complexities when attempting to correlate cosmic occurrences with terrestrial artifacts or geological records. Moreover, the diverse nature and composition of cosmic and terrestrial materials add another layer of complexity to correlation efforts. Identifying commonalities or extraterrestrial influences within terrestrial artifacts requires precise analytical techniques to distinguish between cosmic-origin compounds and naturally occurring terrestrial elements. Resolving these challenges demands an interdisciplinary approach that merges expertise from fields such as astrophysics, geology, archaeology, and materials science. Collaborative efforts to design and implement novel technologies and methodologies will be instrumental in overcoming these obstacles. Establishing standard frameworks for data integration and analysis, as well as fostering open dialogue among researchers from distinct disciplines, will be pivotal in addressing the multifaceted challenges of correlating cosmic and terrestrial data. Looking ahead, advancements in instrumentation and computing capabilities hold promise for enhancing our capacity to probe the cosmic-terrestrial interface, illuminating new pathways for understanding our place within the broader cosmic context.

Future Research Directions

Future research in the field of cosmic anomalies and terrestrial parallels holds considerable promise for advancing our understanding of the universe and its potential impact on Earth. As we navigate the complexities of correlating cosmic and terrestrial data, several crucial avenues for exploration emerge. Firstly, the development of advanced data integration and analysis techniques will be instrumental in identifying subtle correlations and patterns that may elude conventional methodologies. This necessitates collaboration between cosmologists, archaeologists, astrophysicists, and other relevant disciplines to foster interdisciplinary approaches. Moreover, the establishment of dedicated observatories and research centers with a specific focus on anomalous cosmic phenomena can significantly enhance our capacity to monitor, analyze, and interpret potential anomalies at a fundamental level. Understanding these anomalies requires the continued deployment of cutting-edge technological innovations, such as next-generation telescopes, space probes, and high-performance computing systems. Furthermore, the pursuit of international collaborations and data-sharing initiatives is paramount to overcoming the inherent limitations of singular investigations and promoting collective progress in this frontier of exploration. The integration of indigenous knowledge systems and traditional ecological wisdom from diverse cultures can also enrich our perspectives and methodologies, offering holistic insights into anomalous occurrences across different spatial and temporal scales. Additionally, fostering public engagement and education on cosmic anomalies can stimulate interest and participation in citizen science initiatives, thereby broadening the scope of data collection and analysis. Initiatives targeting the involvement of aspiring young researchers and students in anomaly detection and investigation also hold significant potential for nurturing future talent in this intricate field. Lastly, ethical considerations surrounding the exploration of cosmic anomalies and potential impacts on terrestrial systems demand critical evaluation and discourse. Developing frameworks for responsible research conduct, data transparency, and mitigation strategies will be essential as we venture into uncharted territories. By embracing these diverse research directions and adopting a collaborative, multidisciplinary approach, we are poised to unravel enigmatic cosmic anomalies and elucidate their relevance to terrestrial phenomena, unlocking profound insights into the cosmos and our place within it.

Chapter Ten

Exploring Ancient Extraterrestrial Artifacts

The exploration of extraterrestrial artifacts stands at the forefront of contemporary scientific inquiry, offering a profound opportunity to unlock the mysteries of our universe. As humanity continues to probe the boundless frontiers of space and time, the study of ancient extraterrestrial relics provides a unique window into civilizations that may have thrived beyond the confines of our planet. This chapter delves into the historical context, methodologies, and implications of investigating these enigmatic artifacts, shedding light on their potential to reshape our understanding of cosmic history. By examining the remnants of advanced technologies, cultural symbols, and structural formations, researchers seek to piece together the fragmented narratives of distant civilizations that once flourished in the cosmos. The pursuit of extraterrestrial artifact exploration serves as an intellectual bridge between terrestrial archaeology and the burgeoning expanse of cosmic archaeology, uniting two seemingly disparate fields in a shared quest for knowledge. Through this investigative journey, we strive to unravel the intricate tapestry of interstellar influences that may have shaped the evolution of life on Earth and beyond. With each discovery, we are propelled closer to comprehending the interconnectedness of celestial phenomena and human existence, transcending the confines of terrestrial boundaries. This chapter is dedicated to navigating the profound significance of exploring ancient extraterrestrial artifacts and the transformative impact they hold for our comprehension of the cosmic landscape.

Historical Overview of Artifact Discoveries

The study of ancient extraterrestrial artifacts has captivated the human imagination for centuries, providing tantalizing glimpses into the possibility of otherworldly connections. Throughout history, there have been numerous accounts of discoveries that defy conventional explanations and hint at interstellar influences on human civilizations. The earliest documented artifact discoveries date back to ancient Mesopotamia, where intricate carvings and depictions of celestial beings were unearthed, leading scholars to speculate about the existence of beings from beyond our world. Similarly, in ancient Egypt, hieroglyphs and artwork depicted mysterious flying objects and beings that bore striking resemblance to modern-day descriptions of extraterrestrial life. The enigmatic Nazca lines in Peru, created by pre-Columbian civilizations, continue to baffle researchers with their intricate designs and potential astronomical significance. Moving into more recent history, the discovery of the Antikythera mechanism, an ancient Greek analog computer, raises questions about the advanced knowledge and technology possessed by ancient societies. Furthermore, the unexplained presence of metallic objects and formations in various archaeological sites across the globe suggests the existence of sophisticated extraterrestrial craftsmanship or influence. These historical accounts provide a rich tapestry of evidence pointing towards the presence of extraterrestrial artifacts throughout human history. As such, unraveling the historical context and narratives surrounding these enigmatic relics is crucial in understanding the broader implications of extraterrestrial interaction with Earth's inhabitants. By delving into these historical accounts of artifact discoveries, researchers can gain valuable insights into the possible motivations and interactions between ancient civilizations and visitors from the cosmos. This historical overview serves as a foundation for contextualizing the ongoing exploration and analysis of extraterrestrial artifacts, shedding light on the enduring mystery of our cosmic origins and potential encounters with otherworldly beings.

Methodologies for Identifying and Dating Extraterrestrial Objects

The identification and dating of extraterrestrial objects require a multifaceted approach that integrates various scientific disciplines to provide a comprehensive understanding of their origins and significance. Establishing the provenance and age of these artifacts is crucial in unraveling their potential connections to extraterrestrial civilizations. One key methodology for identifying extraterrestrial objects involves thorough morphological and compositional analyses using advanced imaging techniques such as scanning electron microscopy and X-ray fluorescence spectroscopy. These methods allow researchers to examine the structural and elemental composition of the artifacts, providing valuable insights into their manufacturing processes and material sources. Additionally, isotopic analysis using mass spectrometry can help determine the isotopic signatures present in the artifacts, offering clues about their possible extraterrestrial origins. Dating extraterrestrial objects presents unique challenges due to the lack of conventional chronological markers. However, innovative techniques such as luminescence dating, thermoluminescence dating, and optically stimulated luminescence dating can be applied to ascertain the approximate age of these artifacts based on the accumulated radiation-induced energy within their crystalline structures. Furthermore, the integration of geological stratigraphy and contextual analysis can aid in establishing the relative chronology of extraterrestrial objects within archaeological contexts. Corroborating these scientific methodologies with historical and cultural data is essential for contextualizing the significance of the artifacts within the broader framework of extraterrestrial interactions. Interdisciplinary collaboration between archaeologists, astrophysicists, geologists, and other experts is vital in advancing the methodologies for identifying and dating extraterrestrial objects, fostering a holistic approach toward understanding these enigmatic relics. It is imperative to continuously refine and adapt these methodologies as new discoveries emerge, propelling the field of cosmic archaeology into unprecedented frontiers of exploration and insight.

Technological Analysis of Ancient Alien Artifacts

The technological analysis of ancient alien artifacts represents a pivotal aspect of cosmic archaeology, shedding light on the advanced capabilities and knowledge possessed by extraterrestrial civilizations. This sophisticated investigative process involves the application of cutting-edge scientific techniques to unravel the enigma surrounding these enigmatic relics. Utilizing state-of-the-art instruments such as electron microscopes, spectroscopy, X-ray diffraction, and elemental analysis, researchers meticulously examine the composition, structure, and functionality of these otherworldly objects. Through non-destructive testing and high-resolution imaging, scientists endeavor to discern intricate patterns, unconventional materials, and complex mechanisms that defy terrestrial norms. These advanced technologies enable the meticulous documentation and preservation of artifacts while facilitating an in-depth understanding of their origins and purpose. Furthermore, computational modeling and simulation techniques are employed to reconstruct the cultural, technological, and environmental contexts in which these artifacts were created, providing invaluable insights into extraterrestrial societies and their interactions with cosmic environments. Moreover, interdisciplinary collaborations between materials scientists, archaeologists, astrophysicists, and engineers converge in the comprehensive analysis of these artifacts, fostering a holistic approach that integrates diverse expertise and perspectives. The pursuit of technological analysis transcends conventional boundaries, embracing innovation and synergy across scientific domains in unlocking the mysteries of ancient alien artifacts. As we endeavor to decipher the intricacies of these enigmatic relics, the integration of emerging technologies and interdisciplinary methodologies holds the key to unlocking profound revelations about the cosmic heritage of humanity and its enduring connection to celestial forces.

Notable Extraterrestrial Artifacts

In delving into the realm of extraterrestrial artifacts, it is essential to examine and scrutinize case studies that have provided pivotal insights into the possible existence of ancient alien civilizations. Through meticulous archaeological and technological analyses, several remarkable artifacts have been unearthed, offering thought-provoking evidence of interactions with intelligent beings beyond Earth. One such notable artifact is the Klerksdorp Spheres, discovered in the Ottosdal region of South Africa, which has sparked intense debate within the scientific community. These incredibly spherical objects, dated to be over 3 billion years old, exhibit puzzling symmetry and precision, challenging conventional geological explanations. The Betz Mystery Sphere, found in Florida, presents another enigma with its highly polished surface and unusual electromagnetic properties, prompting speculation about its extraterrestrial origins. Furthermore, the Dendera Light Bulb reliefs adorning the walls of the Hathor temple in Egypt have captured the imagination of many scholars, suggesting advanced lighting technology far ahead of its time. The intricate design and depiction of familiar electrical components have led to conjectures about potential contact with an extraterrestrial civilization possessing advanced knowledge. Additionally, the Antikythera Mechanism, a complex geared device recovered from a shipwreck off the coast of Greece, has astounded researchers with its sophisticated astronomical capabilities, posing questions about the involvement of non-human intelligence in its creation. These compelling case studies emphasize the significance of scrutinizing extraordinary archaeological finds that defy conventional explanations, propelling us toward a deeper understanding of humanity's place within the cosmos.

The Role of Artificial Intelligence in Artifact Decipherment

Artificial intelligence (AI) has revolutionized the field of artifact decipherment, offering unprecedented capabilities for analyzing and interpreting ancient extraterrestrial artifacts. Through advanced algorithms and machine learning, AI enables researchers to discern patterns, translate inscriptions, and reconstruct fragmented relics with remarkable accuracy. In the realm of cosmic archaeology, where conventional human-centric methods may fall short due to the complexity and alien nature of these artifacts, AI serves as a powerful tool for unlocking enigmatic secrets of extraterrestrial cultures.

One of the primary strengths of AI lies in its capacity to process vast amounts of data from diverse sources. When applied to the analysis of extraterrestrial artifacts, AI algorithms can swiftly identify correlations between symbols, languages, and material compositions, shedding light on the cultural and technological achievements of ancient extraterrestrial civilizations. By integrating image recognition and linguistic pattern recognition, AI systems can decipher complex hieroglyphs, codices, and encoded messages, providing invaluable insights into the belief systems, knowledge, and societal structure of these enigmatic civilizations.

Furthermore, AI-driven computational models facilitate the cross-referencing of archaeological findings with astrophysical data, enabling researchers to detect astronomical alignments, celestial references, and cosmological symbolism embedded within extraterrestrial artifacts. This multidisciplinary approach, empowered by AI, paves the way for a deeper understanding of how ancient extraterrestrial cultures perceived the cosmos and their place within it. Additionally, AI's ability to simulate the evolution of extraterrestrial languages and dialects over time offers a unique perspective on linguistic shifts and cultural exchanges within these civilizations.

In the realm of artifact restoration and reconstruction, AI algorithms excel at assembling fragmented or deteriorated extraterrestrial relics, reconstructing three-dimensional representations based on fragmentary evidence, and extrapolating missing components through predictive modeling. The application of AI in this domain not only aids in preserving the integrity of these artifacts but also enables researchers to unveil the sophistication of extraterrestrial craftsmanship and technological prowess.

Despite the immense potential that AI holds for artifact decipherment in cosmic archaeology, ethical considerations and responsible use are paramount. Ensuring that AI operates in alignment with established conservation protocols, respects the cultural heritage of extraterrestrial civilizations, and safeguards against potential biases in interpretation is crucial. Additionally, active collaboration between AI specialists, archaeologists, linguists, and cultural experts is essential to harnessing the full capacity of AI in exploring and understanding ancient extraterrestrial artifacts.

Interdisciplinary Approaches to Understanding Extraterrestrial Cultures

Understanding ancient extraterrestrial cultures necessitates an interdisciplinary approach that integrates expertise from archaeology, astronomy, anthropology, linguistics, and other relevant fields. By fostering collaboration between these disciplines, researchers can create a comprehensive framework for deciphering the complexities of extraterrestrial civilizations.

Archaeology plays a key role in this interdisciplinary pursuit, utilizing traditional excavation techniques to unearth extraterrestrial artifacts. These physical remnants offer valuable insights into the technological, artistic, and societal aspects of alien cultures. Meanwhile, astronomers contribute their expertise in cosmology and astroarchaeology, providing a contextual understanding of the planetary systems and cosmic environments that may have influenced extraterrestrial civilizations.

Anthropological perspectives bring a nuanced understanding of social structures, belief systems, and behavioral patterns, enabling scholars to theorize about the cultural norms and customs of these enigmatic societies. Additionally, linguistic analysis becomes essential in decoding extraterrestrial languages or symbolic communication systems that may be inherent in ancient artifacts.

The integration of these diverse fields also requires the development of new technologies and methodologies tailored to the unique challenges posed by extraterrestrial archaeology. High-resolution imaging techniques and advanced spectroscopy enable the detailed analysis of alien relics, shedding light on their composition and potential functions. Furthermore, the utilization of computational modeling and virtual reconstructions allows researchers to simulate extraterrestrial environments and gain a deeper understanding of how these cultures evolved within their respective planetary systems.

Beyond the scientific and technical aspects, ethical considerations play a crucial role in interdisciplinary endeavors related to extraterrestrial cultures. Discussions surrounding cultural appropriation, respect for alien civilizations, and preservation of extraterrestrial heritage become integral components of this holistic approach. Engaging in responsible research practices ensures that the study of ancient extraterrestrial artifacts is conducted with the utmost sensitivity and reverence.

In conclusion, the interdisciplinary nature of understanding extraterrestrial cultures not only expands our knowledge of cosmic history but also fosters a deeper appreciation for the diversity of life in the universe. By uniting various scientific and scholarly disciplines, we pave the way for a more comprehensive and insightful exploration of extraterrestrial civilizations, igniting the imagination and curiosity of humanity as we endeavor to unravel the mysteries of our cosmic neighbors.

Implications of Ancient Extraterrestrial Artifacts for Modern Science

The discovery and study of ancient extraterrestrial artifacts offer profound implications for modern science, spanning various disciplines and opening new avenues for exploration. From archaeology to astronomy, geology to astrobiology, the insights gained from these artifacts significantly impact our understanding of the cosmos and our place within it. One of the most significant implications is the potential rewriting of history. By uncovering evidence of ancient extraterrestrial visits or interactions with Earth, our understanding of human civilization's origin and development may undergo a paradigm shift. This could challenge established chronological timelines and theories, prompting a reevaluation of historical narratives. Furthermore, the technological sophistication exhibited by these artifacts introduces groundbreaking advancements in materials science, metallurgy, and engineering. The study of alien alloys or advanced machinery not only sheds light on extraterrestrial capabilities but also inspires innovation in terrestrial technology. Moreover, the biological implications of extraterrestrial artifacts cannot be overlooked. Analysis of biological residues, genetic markers, or organic structures found in these artifacts could revolutionize our understanding of astrobiology, providing clues to the existence of life beyond Earth. This intersects with ethical considerations regarding potential contamination and biohazards associated with handling extraterrestrial materials. Additionally, the study of ancient extraterrestrial artifacts has the potential to redefine our perception of space exploration and colonization. Understanding how advanced civilizations interacted with their environments and celestial bodies could inform strategies for future human endeavors in space. Furthermore, the insights gained from these artifacts can inform the search for extraterrestrial life and intelligence in the cosmos. By identifying commonalities or differences in cultural expressions found on other worlds, we can refine our expectations and methodologies for detecting signs of alien civilizations. The interdisciplinary nature of this research prompts collaboration across scientific fields, fostering a more holistic approach to understanding our universe. As we continue to explore the implications of ancient extraterrestrial artifacts for modern science, we stand to unlock new frontiers of knowledge, challenging our preconceptions and driving innovation across scientific domains.

Ethical Considerations in Handling Extraterrestrial Materials

Handling extraterrestrial materials presents a unique set of ethical considerations that must be carefully navigated to ensure responsible and respectful treatment of these artifacts. Such materials possess immense scientific and historical importance, and as such, the handling and preservation of these items should be approached with the utmost care and ethical awareness. One key consideration pertains to the potential impact on indigenous cultures and communities. When dealing with extraterrestrial artifacts found in various locations around the world, it is imperative to work collaboratively with local communities and respect their cultural heritage and traditions. This involves engaging in meaningful dialogue, obtaining informed consent, and ensuring that the research and handling of these materials do not infringe upon the rights or beliefs of indigenous peoples. Furthermore, it is crucial to consider the broader societal implications of disclosing information about extraterrestrial artifacts. The release of such findings could have significant cultural, religious, and geopolitical ramifications. Therefore, responsible and transparent communication strategies are essential to manage public perceptions and prevent potential negative consequences from arising. Another ethical dimension revolves around the preservation and protection of extraterrestrial materials. These artifacts represent a finite and irreplaceable resource, and their conservation should be a top priority. This involves establishing stringent protocols for artifact storage, transportation, and analysis to minimize the risk of damage or degradation. Additionally, ethical guidelines must be established for the dissemination of research findings related to extraterrestrial materials to ensure that information is shared in a manner that upholds academic integrity and respects the contributions of all involved parties. Moreover, ethical considerations extend to the potential implications of extraterrestrial material handling on future generations. As custodians of these artifacts, researchers bear a responsibility to ensure that their actions contribute positively to the legacy of these materials and do not compromise opportunities for future exploration and discovery. This may involve developing frameworks for the ethical use of extraterrestrial materials in education and outreach programs, promoting the responsible stewardship of this heritage for the benefit of future generations. In conclusion, navigating the ethical dimensions of handling extraterrestrial materials demands a conscientious and inclusive approach that safeguards the interests of diverse stakeholders, preserves cultural heritage, and contributes to the advancement of scientific knowledge respectfully and responsibly.

Conclusions and Future Directions in Extraterrestrial Artifact Research

The study of ancient extraterrestrial artifacts has not only presented us with profound ethical considerations but also paved the way for exciting future directions in research. As we conclude this comprehensive exploration, it is evident that the implications of these artifacts are significant, urging us to consider the potential impacts on various aspects of science, culture, and society. It is imperative that we carefully navigate these uncharted waters with a deep sense of responsibility and respect for the potentially groundbreaking discoveries that await us. Reflecting on the ethical considerations discussed in the previous section, it is crucial to emphasize the need for stringent protocols and guidelines governing the handling, preservation, and dissemination of any discovered extraterrestrial materials. This includes ensuring the protection of these artifacts from exploitation and unauthorized access, while also promoting international collaboration and cultural sensitivity in their study and interpretation. Looking towards the future, the field of extraterrestrial artifact research presents numerous compelling avenues for exploration and discovery. Advancements in technology, such as improved analytical methods, remote sensing capabilities, and artificial intelligence applications, hold promise for enhancing our understanding of these enigmatic relics. Additionally, interdisciplinary collaborations between cosmologists, archaeologists, anthropologists, and other related fields can provide unique perspectives and contribute to a more holistic comprehension of ancient extraterrestrial cultures and their influence on human history. Furthermore, the establishment of rigorous academic programs and research initiatives dedicated to the study of cosmic archaeology can foster a new generation of scholars and experts equipped to tackle the challenges and opportunities presented by extraterrestrial artifact research. Considering the broader societal impact, the potential identification of verifiable extraterrestrial artifacts could revolutionize our understanding of humanity's place in the universe, raising thought-provoking questions about our cosmic origins and the existence of other intelligent civilizations. Addressing these profound questions requires careful consideration of the philosophical, ethical, and existential implications that may arise, underscoring the necessity for open dialogue and critical reflection within the global community. In conclusion, as we chart the course for future research in the realm of ancient extraterrestrial artifacts, it is clear that we stand at the precipice of a transformative era in human knowledge and understanding. By upholding ethical standards, fostering interdisciplinary collaboration, and embracing the uncertainties and possibilities inherent in this pursuit, we can embark on a journey that not only enriches our scholarly pursuits but also shapes our collective vision of the cosmos and our place within it.

Chapter Eleven

SETI and Archaeology: A Synergistic Approach

The convergence of SETI methodologies with archaeological practices marks an unprecedented step towards unifying the quest for understanding our place in the cosmos. This integrated approach seeks to leverage the tools, techniques, and perspectives developed within the domains of Search for Extraterrestrial Intelligence (SETI) and archaeology, recognizing the potential shared goals and interests between these traditionally distinct scientific disciplines. By exploring the historical overview of SETI and archaeological methods, it becomes evident that both fields share a common curiosity about the origins, evolution, and potential diversity of life in the universe. The overarching goal of this integrated approach is to harness the collective knowledge, expertise, and methodologies from these fields to unravel the mysteries surrounding extraterrestrial intelligence and its potential manifestations within archaeological contexts. Through an interdisciplinary lens, the shared pursuit of understanding the cosmos not only deepens our exploration of potential extraterrestrial signals but also enriches our comprehension of human history, societal evolution, and cultural interactions. As such, this collaborative endeavor represents a paradigm shift in how we perceive and investigate the cosmic enigma, fostering a holistic and interconnected approach that transcends the boundaries of traditional academic silos. The introduction of advanced technological tools and analytical frameworks further underlines the symbiotic relationship between SETI and archaeology, enabling synergistic investigations that go beyond conventional methodological confines. By embracing this unified approach, researchers aim to unlock new insights into the possibilities of extraterrestrial intelligence while simultaneously shedding light on the complexities of ancient civilizations, material culture, and their potential interactions with cosmic phenomena. The integration of artificial intelligence and machine learning algorithms serves as a catalyst for comprehensive data analyses and pattern recognition, bridging the gap between disparate datasets to reveal nuanced connections between terrestrial and celestial phenomena. This chapter sets the stage for a profound exploration of how the convergence of SETI and archaeology opens unparalleled avenues for discovery, dispelling the boundaries between Earth-bound anthropocentrism and cosmic perspectives, and ultimately deepening our understanding of the intricate tapestry woven by humanity and the cosmos.

Historical Overview of SETI and Archaeological Methods

The historical roots of the search for extraterrestrial intelligence (SETI) and the development of archaeological methods to explore ancient civilizations share a common thread of intellectual curiosity and a quest for knowledge. Dating back to the early 20th century, the idea of actively seeking evidence of intelligent life beyond Earth captivated the minds of scientists and visionaries. Meanwhile, the field of archaeology has evolved from an antiquarian pursuit to a multidisciplinary science focusing on unraveling the complexities of human history and culture.

Early proponents of SETI, such as Nobel laureate Frank Drake, laid the groundwork for systematic searches for signals from advanced civilizations. Concurrently, pioneers in the field of archaeology, including

William Flinders Petrie and Kathleen Kenyon, revolutionized methodologies for excavating ancient sites and interpreting material culture.

The 20th century witnessed significant advancements in both fields. The advent of radio astronomy provided SETI researchers with powerful tools for scanning the cosmos, culminating in the establishment of programs like the Search for Extraterrestrial Intelligence Institute (SETI Institute) and initiatives such as Breakthrough Listen. Similarly, archaeologists leveraged technological innovations such as ground-penetrating radar, LiDAR, and advanced imaging techniques to uncover hidden structures and artifacts, shedding light on the enigmatic aspects of human civilization.

Throughout history, the intersection between SETI and archaeology has led to intriguing parallels. Both disciplines grapple with deciphering cryptic messages from the past, whether encoded in ancient inscriptions or cosmic signals. Moreover, they share a common goal: piecing together fragments of evidence to reconstruct narratives of civilizations, be they terrestrial or extraterrestrial.

The synergy between SETI and archaeology is further underscored by the growing recognition that they can inform and enrich each other's methodologies. As archaeologists adapt sophisticated data analysis techniques from SETI for pattern recognition and anomaly detection, SETI researchers draw inspiration from archaeological approaches to understand the societal and environmental factors that may shape extraterrestrial civilizations.

This historical overview sets the stage for a deeper exploration of the interconnectedness between SETI and archaeology, laying the groundwork for a synergistic approach that transcends traditional disciplinary boundaries.

Technological Tools: Bridging the Methodological Gap

As we delve into the realm of unified investigations between SETI and archaeology, it becomes increasingly evident that technological advancements play a pivotal role in bridging the methodological gap. The intersection of these disciplines necessitates the development and implementation of cutting-edge tools that can cater to the unique demands of both fields. In this chapter, we will explore the diverse array of technological tools that have been instrumental in facilitating collaborative research endeavors between SETI and archaeology. Technologies such as remote sensing devices, advanced imaging systems, spectroscopic methods, and geological mapping instruments have proven indispensable in uncovering extraterrestrial artifacts and deciphering cryptic archaeological enigmas. These innovative tools not only enhance the efficiency of data collection but also contribute to the precision and accuracy of our analyses. Furthermore, the integration of computational modeling and simulation software has revolutionized the visualization and interpretation of complex archaeological and SETI datasets. By harnessing the power of machine learning algorithms, researchers can now extract intricate patterns from vast repositories of data, shedding light on elusive connections that may have previously eluded detection. Moreover, the burgeoning field of drone technology has opened up new frontiers in archaeological surveying and exploration, enabling us to access remote and inaccessible sites with unparalleled ease and precision. The seamless fusion of these technological marvels serves to blur the boundaries between traditional methodologies, forging a cohesive framework for synergistic investigations. It is imperative to acknowledge that the rapid pace of technological innovation presents both opportunities and challenges for researchers seeking to converge the realms of SETI and archaeology. As we navigate this dynamic landscape, the ethical considerations surrounding the use of advanced technologies in archaeological and extraterrestrial research cannot be overstated. Striking a balance between technological reliance and the preservation of cultural heritage is paramount in ensuring the responsible and sustainable advancement of these interdisciplinary pursuits. With a keen focus on leveraging state-of-the-art technologies to unravel the mysteries of cosmic antiquity, the collaborative efforts of SETI and archaeology are poised to redefine our understanding of humanity's place in the cosmos.

Synergistic Investigations in Practice

In the burgeoning field of Cosmic Archaeology, the application of synergistic approaches involving both SETI and traditional archaeological methods has yielded fascinating case studies. One such notable investigation involved the joint analysis of anomalous cosmic signals alongside the examination of ancient petroglyphs in a remote archaeological site. The synchrony between these two seemingly distinct areas of study revealed compelling parallels that could not be overlooked. Through this case study, researchers were able to discern potential correlations in the symbols depicted in the petroglyphs with the patterns observed in the cosmic signals, leading to an intriguing hypothesis that ancient civilizations might have had awareness of celestial phenomena beyond what was previously postulated. Another captivating case study revolved around the strategic integration of satellite imaging data obtained from space exploration missions with ground-based archaeological surveys. This holistic approach allowed for the identification of anomalous geological formations that aligned with cosmic impact events, shedding light on the interconnectedness of cosmic events and their potential influence on ancient cultures. Moreover, the collaborative efforts between SETI specialists and archaeologists in analyzing historical narratives and cultural artifacts led to the recognition of recurring motifs that mirrored specific celestial occurrences. These profound insights have fueled a paradigm shift in our understanding of ancient civilizations and their possible interactions with extraterrestrial forces, offering fresh perspectives on humanity's cosmic origins. Furthermore, the application of high-resolution spectroscopic techniques traditionally used in astrobiology to analyze organic residues extracted from ancient artifacts has provided a novel avenue for investigating potential extraterrestrial biomarkers within archaeological contexts. These interdisciplinary investigations have not only broadened our methodological repertoire but have also instigated a nuanced discourse on the potential dissemination of extraterrestrial organic matter onto Earth. The intricate nature of these case studies underscores the significance of embracing a synergistic approach, illustrating the depth of knowledge that can be unearthed through the convergence of SETI and Archaeology.

Data Analysis Techniques from SETI Applied to Archaeology

In the field of cosmic archaeology, the application of data analysis techniques from the search for extraterrestrial intelligence (SETI) has opened up new frontiers for archaeological research. Building upon the advanced computational methodologies developed in the pursuit of identifying potential intelligent signals from distant civilizations, archaeologists have begun to explore how these techniques can be adapted and utilized in the analysis of ancient artifacts and cultural patterns on Earth. This innovative approach involves integrating signal processing algorithms, machine learning, and statistical modeling into traditional archaeological practices. By doing so, researchers have significantly expanded their ability to identify, document, and interpret traces of ancient technological and cultural activities. One key area of crossover is the utilization of pattern recognition algorithms originally developed for analyzing radio signals in the identification of complex patterns within archaeological data. These algorithms are proving invaluable in detecting subtle correlations and recurrent motifs across different archaeological sites, aiding in the identification of potential cultural exchange networks and technological innovations. Furthermore, the application of machine learning models allows for the classification and predictive analysis of archaeological findings, shedding light on the possible functions and contexts of enigmatic artifacts. Another crucial intersection lies in the realm of statistical analysis and modeling. Archaeologists are leveraging probabilistic methods and data-driven approaches used in SETI to assess the significance of archaeological discoveries, quantify uncertainties, and reconstruct past behaviors with greater precision. By incorporating these advanced data analysis techniques, the field of cosmic archaeology stands on the brink of transformative breakthroughs. However, this interdisciplinary endeavor also poses significant challenges. The translation of highly specialized algorithms and computational methodologies from SETI into the archaeological domain requires careful calibration and validation to ensure their reliability in diverse archaeological contexts. Moreover, ethical considerations surrounding data interpretation and the potential emergence of false positives demand vigilant scrutiny in the application of these techniques. As the synergy between SETI and archaeology evolves, it holds immense promise for unraveling the mysteries of ancient civilizations and shedding light on our place in the cosmos.

Archaeological Insights into SETI Research

As we delve deeper into the synergistic approach of combining SETI and archaeology, it becomes apparent that the marriage of these two disciplines offers a wealth of insights and potential discoveries. While the primary focus of SETI has been on detecting extraterrestrial intelligence through radio signals and other technological methods, the archaeological perspective brings a unique dimension to the table. By examining ancient artifacts, historical sites, and cultural remnants, archaeologists can offer valuable insights into the potential manifestations of extraterrestrial influence on Earth.

One compelling area of exploration lies in the analysis of ancient structures and artifacts for anomalous features or materials that may suggest extraterrestrial origins. The careful study of ancient civilizations' depictions of celestial phenomena, unidentified flying objects, or encounters with non-human entities can provide clues that complement traditional SETI data. Moreover, the comparison of cultural narratives and mythologies across diverse human societies might converge on common themes that could indicate interactions with extraterrestrial beings.

Furthermore, the investigation of archaeological sites within potential astrobiological hotspots, such as Mars or Europa, presents a tantalizing prospect for uncovering evidence of past or present extraterrestrial life. By leveraging geological and biological expertise along with advanced imaging technologies, archaeologists equipped with a cosmic lens could contribute significantly to future SETI missions and planetary explorations.

Another pivotal aspect involves the synthesis of cultural anthropology with the search for extraterrestrial intelligence. Cultural practices and rituals related to astronomical events, star worship, or cosmic symbolism may hold hidden meanings that intersect with the quest for understanding potential extraterrestrial communications or transmissions.

Unraveling the enigma of cosmic anomalies and artifacts discovered on Earth also holds promise in shedding light on the broader quest for cosmic knowledge. The rigorous application of archaeological methodologies, such as stratigraphy, typology, and contextual analysis, can yield invaluable context to materials of potential extraterrestrial origin, supporting the meticulous deciphering of their significance in the grander tapestry of universal heritage.

Ultimately, the collaboration between SETI and archaeology opens up new vistas of inquiry and discovery, bridging the cosmic and terrestrial realms in pursuit of a deeper understanding of our place in the universe. As we continue to explore these interdisciplinary frontiers, the potential for profound revelations awaits us at the intersection of past and future, human and extraterrestrial, archaeology and SETI.

Interdisciplinary Challenges and Solutions

Interdisciplinary research at the intersection of SETI and archaeology presents a myriad of challenges that require innovative solutions. The fusion of two distinct fields necessitates overcoming inherent disparities in methodologies, terminologies, and epistemologies. One of the primary challenges lies in establishing a common ground for communication and collaboration between experts in both domains. This involves developing a shared framework of understanding that integrates concepts from astrobiology, anthropology, astronomy, and advanced technological disciplines.

Furthermore, the complex nature of data integration and interpretation poses a significant obstacle. Combining massive volumes of astronomical observations with intricate archaeological findings demands robust computational models and analytical tools. As diverse datasets merge, reconciling contrasting scales, resolutions, and uncertainties becomes a daunting task in itself.

Another pervasive challenge revolves around funding and institutional support. Interdisciplinary research often faces barriers in resource allocation due to its non-traditional nature. Securing sustainable financing for unified research endeavors and garnering institutional recognition remains an arduous endeavor.

Moreover, the ethical dimensions of conducting interdisciplinary research hold paramount significance. Delving into potential extraterrestrial artifacts and civilizations calls for careful consideration of cultural, societal, and environmental implications. Negotiating ethical frameworks and engaging relevant stakeholders is essential to navigate these uncharted territories responsibly.

In addressing these challenges, several innovative solutions can be envisaged. Firstly, fostering cross-disciplinary education and training programs is imperative to cultivate a new generation of researchers equipped with versatile skill sets. By nurturing hybrid expertise, future scholars can bridge the gap between SETI and archaeology more seamlessly.

Additionally, establishing dedicated interdisciplinary research centers and consortiums can serve as focal points for collaboration, knowledge exchange, and advocacy. These initiatives can provide crucial infrastructure and networking opportunities, thereby nurturing a vibrant community of interdisciplinary scholars.

Development of advanced machine learning algorithms tailored for analyzing heterogeneous datasets promises to revolutionize integrated research methodologies. Leveraging artificial intelligence and big data analytics enables automated pattern recognition, outlier detection, and algorithmic inference across disparate data types.

Lastly, advocating for policy reforms and international cooperation is vital to create a conducive environment for interdisciplinary research. Engaging with governmental bodies, scientific agencies, and public institutions to champion the value of uniting SETI and archaeology can pave the way for enhanced support and recognition.

By navigating these challenges proactively and embracing innovative solutions, the synergy between SETI and archaeology can evolve into a harmonized frontier of discovery, transcending conventional disciplinary boundaries.

The Role of Artificial Intelligence in Unified Analyses

Artificial Intelligence (AI) has become an indispensable tool in the field of unified analyses, where it serves to bridge the gap between data from SETI and archaeological investigations. AI can process vast amounts of complex data, identify patterns, and make connections that would be challenging for human researchers alone. In the context of SETI and archaeology, AI can play a pivotal role in identifying potential signals or artifacts that may not be immediately apparent to human observers. By employing advanced algorithms and machine learning techniques, AI can sift through large datasets derived from both disciplines and extract meaningful insights that could lead to groundbreaking discoveries. One of the main advantages of using AI in unified analyses is its ability to discover patterns and translate the unknown into tangible knowledge. Through sophisticated data mining and analysis, AI can uncover correlations and anomalies that may have previously eluded researchers, shedding new light on historical and cosmic phenomena. Additionally, AI can help automate repetitive tasks, freeing up researchers' time to focus on more complex analyses and interpretation. With the aid of AI, researchers can explore uncharted territories within the realms of SETI and archaeology, opening doors to new avenues of research and discovery. Furthermore, as technology continues to advance, AI holds the potential to revolutionize the way data from both disciplines is collected, processed, and interpreted. The integration of AI systems with innovative sensing technologies could enhance the efficiency and accuracy of data collection, leading to more comprehensive and reliable datasets. Moreover, AI-driven predictive modeling can assist in simulating various scenarios, allowing researchers to anticipate potential outcomes and plan targeted investigations. As a result, the role of AI in unified analyses extends beyond data processing; it encompasses the facilitation of predictive analytics and decision support, empowering researchers to make informed choices and maximize the impact of their investigations. Looking ahead, the continued development of AI holds promise for unlocking the mysteries of the cosmos and our ancient past. By leveraging the capabilities of AI in unified analyses, researchers can harness the power of technology to unravel enigmas that have long evaded conventional methodologies. As we venture into the future, AI stands as a formidable ally in the quest for knowledge and understanding, poised to revolutionize the synergistic approach of SETI and archaeology.

Future Directions and Potential Breakthroughs

As we stand at the intersection of SETI and archaeology, contemplating the fusion of cosmic exploration and terrestrial investigation through the lens of artificial intelligence, it is imperative to recognize the vast potential for future breakthroughs in this dynamic field. The convergence of advanced technologies with the timeless pursuit of unraveling cosmic mysteries holds promise for transformative insights that could redefine our understanding of our place in the universe. One avenue of future exploration lies in the harnessing of machine learning algorithms to sift through massive datasets, both terrestrial and extraterrestrial, enabling the swift identification of patterns, anomalies, and potential signals indicative of intelligent extraterrestrial life or ancient advanced civilizations. Moreover, with developments in quantum computing on the horizon, the ability to process and analyze unprecedented volumes of data could propel the synergistic approach of SETI and archaeology into uncharted territory. This technological leap not only opens avenues for more nuanced analysis of existing artifacts and celestial data but also sets the stage for the discovery of entirely new classes of cosmic phenomena or archaeological evidence. Furthermore, the burgeoning field of astrobiology offers a compelling direction for future research, forging interdisciplinary collaborations between biologists, chemists, astronomers, and archaeologists to investigate the potential for life beyond Earth and its implications for our understanding of ancient cosmic civilizations. As we delve deeper into the realms of astroarchaeology, the development of innovative instrumentation and observation techniques becomes paramount. Advanced telescopic systems, incorporating multi-spectral imaging and adaptive optics, hold great promise for detecting elusive traces of ancient extraterrestrial endeavors, potentially unveiling archaeological remnants within our solar system and beyond. The potential integration of space missions explicitly designed to search for alien artifacts further underscores the compelling prospects for future breakthroughs in this burgeoning field. Moreover, future directions must consider the ethical implications of discovering evidence of advanced extraterrestrial civilizations, prompting the need for robust frameworks ensuring responsible and coordinated global responses. Additionally, the development of effective communication strategies to convey complex scientific findings to diverse audiences becomes crucial, fostering public engagement and support for continued explorations. Embracing an interdisciplinary and forward-thinking approach, the future melding of SETI and archaeology holds the promise of reshaping our cosmic narrative and may ultimately pave the way for profound shifts in our perception of the cosmos and our place within it.

Summary and Implications for Future Research

In summarizing the synergistic approach of SETI and archaeology, it is evident that this combined field holds great potential for unlocking profound revelations about our place in the cosmos. The ongoing advancement in technological capabilities, data analysis tools, and cross-disciplinary collaboration has paved the way for an exciting future of research. As we look ahead, it becomes increasingly imperative to recognize the implications and potential impact of further investigations in this domain.

The intersection of SETI and archaeology presents a unique opportunity to address fundamental questions about extraterrestrial life and intelligence, as well as our connection to other civilizations beyond Earth. The integration of historical perspectives and methodological advancements from both fields offers a comprehensive approach to understanding humanity's origins and our potential cosmic neighbors.

One of the key implications for future research lies in the development and deployment of advanced AI systems to aid in unified analyses of SETI and archaeological data. The use of AI can revolutionize the field by automating complex pattern recognition, enhancing data processing efficiency, and revealing correlations that may elude human perception. Furthermore, the ethical considerations surrounding AI utilization in deciphering potential extraterrestrial artifacts warrant careful deliberation and oversight.

In addition to harnessing AI capabilities, the collaborative efforts of diverse experts, including astrophysicists, archaeologists, data scientists, and cultural historians, will be pivotal for progressing the synergistic approach. Interdisciplinary challenges such as establishing common frameworks for data interpretation and fostering effective communication across specialized domains must be addressed to fully leverage the collective expertise for groundbreaking discoveries.

Moreover, the burgeoning advancements in space exploration technologies offer promising avenues for expanding the search for cosmic echoes. The incorporation of innovative instruments, such as next-generation telescopes, deep-space probes, and unmanned missions to celestial bodies, can significantly enhance the capacity to detect potential traces of ancient extraterrestrial civilizations and artifacts.

As we pursue these future research endeavors, close attention must be dedicated to maintaining scientific rigor and transparency. This includes rigorous peer review processes, replicable analyses, and transparent reporting of findings. Ensuring the reliability and reproducibility of results is essential for upholding the credibility of the growing body of evidence in the field of cosmic archaeology.

Finally, the implications for future research extend beyond the realms of scientific inquiry, delving into philosophical and societal dimensions. The potential discovery of evidence pertaining to extraterrestrial civilizations would undoubtedly prompt profound contemplation on humanity's place in the universe and our ethical responsibilities towards alien entities if they exist. Addressing these profound questions requires thoughtful reflection and inclusive dialogue among global stakeholders.

In essence, the convergence of SETI and archaeology opens vistas to explore age-old mysteries and unfold new chapters in our civilization's narrative. The quest for cosmic truths beckons us to embark on a journey that transcends disciplinary boundaries and invites us to contemplate our significance in a universe teamed with possibilities. The horizon of cosmic archaeology stands as a testament to humanity's enduring curiosity and resilience in the face of the unknown, inspiring us to venture forth into uncharted territories of knowledge and discovery.

Chapter Twelve

Unveiling Alien Civilizations

The proposition of civilizations beyond Earth has captivated human imagination for centuries, fueling an enduring curiosity about the existence of life elsewhere in the cosmos. As we stand on the threshold of unprecedented advancements in space exploration and the search for extraterrestrial intelligence, the subject of extraterrestrial civilizations has garnered renewed attention from scientists, researchers, and the general public alike. The idea of advanced societies inhabiting distant exoplanets or even within our own solar system presents a tantalizing prospect, prompting us to ponder profound questions about the nature of intelligence, the evolution of technological capabilities, and the potential diversity of cultural expressions across the cosmos. This chapter seeks to delve into the complexities and implications of contemplating civilizations that may lie beyond the confines of Earth, laying the groundwork for an exploration of historical precedents, theoretical frameworks, and the multifaceted dimensions of this revelatory pursuit. By probing the origins of our fascination with the alien other and surveying the spectrum of conjectures about extraterrestrial societies, we aim to foster a deeper understanding of the significance and implications of this compelling intellectual endeavor.

Historical Context and Theoretical Framework

Throughout human history, the concept of extraterrestrial civilizations has captivated the imagination of scholars, scientists, and enthusiasts. This fascination can be traced back to ancient civilizations such as the Sumerians, Egyptians, and Maya, whose mythologies and cosmologies often incorporated narratives of interactions with otherworldly beings or advanced celestial creatures. These early accounts serve as a testament to humanity's enduring curiosity about the possibility of life beyond Earth.

The modern exploration of the historical context and theoretical framework of alien civilizations is deeply rooted in the pioneering work of visionaries like Giordano Bruno, who in the 16th century proposed the existence of other worlds and speculated on the potential attributes of inhabitants that might dwell there. Over time, the field has evolved, drawing from diverse disciplines such as astronomy, anthropology, sociology, and physics to construct a comprehensive framework for investigating the hypothetical existence of extraterrestrial societies. This interdisciplinary approach seeks to synthesize historical evidence, scientific conjecture, and theoretical postulations to provide a multifaceted understanding of the prospect of alien civilizations.

Central to the theoretical framework are conceptual models and thought experiments that attempt to contextualize the societal dynamics and technological achievements of putative extraterrestrial cultures. Researchers have deliberated on topics ranging from the evolution of alien communication systems to the possible structures of their sociopolitical organizations, leveraging insights from terrestrial analogies and extrapolating them onto interstellar scales. The speculative nature of this endeavor is tempered by scholarly rigor, with an emphasis on empirical observations and logical deductions.

Furthermore, the historical context and theoretical framework encompass an array of influential hypotheses, including the Fermi paradox, the Drake equation, and the Kardashev scale, each offering distinct lenses through which to analyze and interpret potential signals or traces of extraterrestrial intelligence. These frameworks serve as intellectual constructs that enable researchers to navigate the complexities intrinsic to the investigation of alien civilizations while also fostering critical discourse and theoretical refinement within the scientific community.

Ultimately, delving into the historical antecedents and underpinning theories of alien civilizations engenders a profound appreciation for humanity's place within the cosmic tapestry, prompting reflection on our existential interconnectedness with potential interstellar counterparts and inviting contemplation on the broader implications of contact with an advanced extraterrestrial society.

Technological Signatures: Detection and Analysis

The search for technological signatures of extraterrestrial civilizations is a complex and multi-faceted endeavor that requires meticulous detection and comprehensive analysis. In this quest, scientists and researchers employ cutting-edge technologies and innovative methodologies to identify potential indicators of advanced alien technology within the cosmos. The primary focus lies in scrutinizing celestial phenomena for atypical patterns, energy emissions, or artificial signals that deviate from natural processes. This involves leveraging sophisticated telescopes, spectroscopy, and radio telescopes to peer into the depths of space and unravel any enigmatic anomalies that could suggest the presence of advanced extraterrestrial civilizations.

Detection of technological signatures also encompasses the scrutiny of exoplanetary systems for irregularities, such as unusual planetary configurations, megastructures, or artificial transit signals. Innovative data analysis techniques, including machine learning algorithms and data mining strategies, are integral tools in parsing through vast datasets obtained from space observatories and telescopic surveys. By discerning potential artificial modulations or structured patterns within these datasets, researchers can unearth potential indicators of advanced technological activity that may offer compelling evidence of extraterrestrial intelligence.

Moreover, the analysis of cosmic phenomena extends beyond astronomical observations, encompassing the investigation of cosmic microwave background radiation and anomalous cosmic phenomena. Scientists meticulously sift through colossal datasets to discern any atypical signals or discernible signatures of advanced technology that may be embedded within the cosmic fabric. Furthermore, the pursuit of unconventional cosmic artifacts or relics holds promise in uncovering remnants of ancient alien technology embedded within the celestial sphere, providing insights into the technological prowess of extraterrestrial civilizations.

Researchers must adopt a multidisciplinary approach, integrating diverse fields such as astrophysics, astrobiology, and information theory to comprehensively analyze and interpret potential technological signatures. Collaborative endeavors among scientific communities on a global scale are vital to harnessing collective expertise and resources in scrutinizing the vast cosmic landscape for hints of extraterrestrial technological artifacts and activities. Moreover, ethical considerations and rigorous validation protocols are intrinsic components of this pursuit, ensuring the veracity and significance of any purported technological signatures attributed to alien civilizations.

In essence, the quest to detect and analyze technological signatures of extraterrestrial civilizations stands as a testament to humanity's insatiable curiosity and unwavering pursuit of knowledge about our place in the universe.

Sociocultural Structures of Alien Societies

Understanding the sociocultural structures of alien societies is a complex yet fundamental aspect of cosmic archaeology. In delving into this subject, we are confronted with the intricate web of social organization, belief systems, and cultural practices that may exist in extraterrestrial civilizations. While the study of alien cultures is inherently speculative, it is not without merit in expanding our understanding of the universe and our place within it.

In examining the sociocultural structures of potential alien societies, we must consider the concept of social hierarchy, governance systems, and the distribution of power. It is essential to contemplate the various models of societal organization that could exist beyond our planet, ranging from hierarchical to egalitarian structures. Exploring possible forms of governance, decision-making processes, and social stratification allows us to envision the diversity of alien societies and challenge our own preconceptions rooted in earthly experiences.

Moreover, the examination of belief systems and cultural practices is paramount in deciphering the social fabric of alien civilizations. Religious and spiritual traditions, mythologies, rites, and rituals may shed light on the values, worldview, and existential concerns of extraterrestrial beings. The comparative analysis of these aspects with terrestrial cultures can offer intriguing parallels or stark contrasts, prompting introspection into the nature of human belief systems and their evolution over time.

Another crucial facet of sociocultural exploration is language and communication within alien societies. The study of extraterrestrial linguistics encompasses not only deciphering symbolic communication but also comprehending non-verbal modes of expression. Understanding the linguistic foundations of an alien civilization can offer profound insights into their cognitive processes, social interactions, and the transmission of knowledge across generations.

As we navigate through the complexities of sociocultural structures in the search for potential extraterrestrial civilizations, we must approach this endeavor with humility, open-mindedness, and interdisciplinary collaboration. The implications of encountering diverse alien societies extend far beyond the realms of archaeology and astrophysics, permeating into the domains of philosophy, ethics, and interstellar diplomacy. By engaging in this exploration, we broaden our perspectives, enrich our intellectual discourse, and lay the groundwork for meaningful dialogues about our cosmic neighbors.

Linguistic Considerations in Decoding Communication

In the pursuit of understanding extraterrestrial civilizations, linguistic considerations play a vital role in decoding communication patterns. The challenge lies in deciphering languages and forms of communication that may be fundamentally different from anything encountered on Earth. This requires an interdisciplinary approach, drawing insights from linguistics, cryptography, and information theory.

Language itself is a complex web of meaning, context, and structure. When approaching the study of alien languages, researchers must grapple with the possibility of entirely novel semantic frameworks, grammatical structures, and phonological systems. In some cases, a mode of communication might not rely on sound at all, but instead, on visual or tactile modalities. The absence of a shared evolutionary history could mean that extraterrestrial languages are not bound by the same cognitive or physiological constraints as human languages, posing significant challenges for decoding.

Moreover, the cultural and social contexts shaping language use among alien civilizations remain a matter of speculation. These intricate aspects of communication underscore the need for a nuanced examination of the cultural underpinnings, social dynamics, and historical evolution of extraterrestrial societies. Linguistic anthropologists, computational linguists, and semioticians contribute valuable perspectives to this complex task, recognizing that language serves as a mirror of cultural identity, societal norms, and shared knowledge within a community.

The study of linguistic universals and the development of a framework for comparative linguistic analysis are imperative in identifying commonalities that could provide a starting point for understanding alien communication. Semantic primes, syntax, and pragmatics are essential components to explore when attempting to derive meaning from alien linguistic data. Furthermore, the potential existence of non-verbal communication, such as gestural or olfactory modes, introduces additional layers of complexity to the decoding process.

Interdisciplinary collaboration becomes instrumental in unraveling the intricacies of extraterrestrial communication. Advanced computational algorithms, machine learning models, and neural networks would aid in processing vast datasets of linguistic information, while cognitive scientists and neurologists work towards understanding the cognitive and perceptual mechanisms underlying alien communication. However, the ethical considerations surrounding the interrogation of alien languages and the ramifications of potential misinterpretation cannot be overlooked.

Ultimately, the endeavor to decode extraterrestrial languages necessitates a profound reevaluation of our preconceived notions about communication, cognition, and culture. As we venture into the uncharted territories of interstellar linguistics, we must remain vigilant in our approach, respecting the inherent diversity and complexity that may characterize communication beyond our world.

Archaeological Sites and Their Cosmic Implications

The search for evidence of extraterrestrial civilizations extends to the investigation of archaeological sites with potential cosmic implications. Archaeology, as a discipline, has traditionally been focused on terrestrial artifacts and the study of human history. However, with advancements in space exploration and astroarchaeology, there is growing interest in exploring sites that may hold clues to interactions with extraterrestrial entities.

Archaeological sites with cosmic implications can manifest in various forms, ranging from ancient structures that exhibit advanced technological features beyond the capability of the known human civilization to enigmatic artifacts with potential extraterrestrial origins. These sites pose intriguing questions about the possibility of past contact or influence by advanced cosmic civilizations. The meticulous excavation and analysis of such sites present unique challenges, including the integration of multidisciplinary approaches from archaeology, astrobiology, and exoarchaeology.

One of the primary objectives of studying these archaeological sites is to discern any patterns or anomalies that deviate from the expected development of human societies. This involves employing cutting-edge technologies, such as remote sensing, spatial imaging, and spectroscopic analysis, to unearth hidden complexities within the structures and artifacts. Additionally, the examination of isotopic compositions and trace elements enables researchers to assess the potential off-world origins of materials, further unraveling the cosmic implications of the site.

Furthermore, the interpretation of archaeological findings within a cosmic context necessitates a reevaluation of established cultural and historical narratives. It calls for a critical reassessment of conventional belief systems and societal evolution models to accommodate the prospect of extraterrestrial intervention. The collaborative efforts of expert linguists, anthropologists, and historians become pivotal in deciphering the underlying meanings embedded in linguistic scripts, symbols, and inscriptions found at these sites, possibly shedding light on interstellar communication and interactions.

The significance of investigating archaeological sites with cosmic implications extends beyond scientific curiosity; it delves into existential inquiries regarding humanity's place in the cosmos. By unraveling the enigmas concealed within these sites, we advance our understanding of the potential interconnectedness between terrestrial and extraterrestrial civilizations, offering profound insights into the broader narrative of cosmic evolution and the implications for humanity's future.

Interstellar Diplomacy: Ethical and Political Considerations

Interstellar diplomacy presents a complex and nuanced set of ethical and political considerations that extend far beyond the terrestrial realm. As we contemplate the potential existence of extraterrestrial civilizations, we are compelled to consider the implications of initiating contact and establishing diplomatic relations with entities whose sociocultural structures, technological advancements, and ethical frameworks remain wholly unfamiliar to us.

The ethical dimensions of interstellar diplomacy revolve around the fundamental respect for the autonomy and sovereignty of any discovered alien civilizations. This necessitates careful deliberation on the potential impact of our actions on their societal structures, belief systems, and overall well-being. Moreover, questions surrounding the equitable exchange of knowledge, resources, and technology must be approached with utmost sensitivity to avoid inadvertently causing harm or disruption to extraterrestrial societies.

In parallel, the political challenges inherent in interstellar diplomacy cannot be understated. The encounter with an advanced alien civilization could potentially trigger geopolitical tensions, power struggles, and conflicting global agendas on Earth. It becomes imperative to establish international protocols and institutions dedicated to managing the intricacies of interstellar engagement, ensuring transparency, and averting the misuse of newfound cosmic knowledge for political gain or military purposes.

Furthermore, the question of representation and inclusivity in interstellar negotiations demands thorough examination. How do we ensure that diverse voices from Earth, including marginalized communities and developing nations, have a fair say in shaping our approach to extraterrestrial diplomacy? This highlights the necessity of fostering equity and preserving cultural diversity in our interactions with hypothetical alien civilizations.

The establishment of ethical codes and guidelines governing interstellar diplomacy represents an indispensable endeavor. It requires the collaboration of scholars, policymakers, ethicists, and representatives from various fields to craft a framework that upholds universal values while remaining flexible enough to accommodate the unforeseen complexities of such momentous encounters.

In the pursuit of interstellar diplomacy, exemplifying humility, empathy, and unwavering commitment to ethical conduct becomes paramount. While navigating the uncharted territories of cosmic relations, humanity stands poised to redefine its place in the cosmos as responsible stewards of not only our planet but potentially the broader interstellar community.

Comparative Analysis of Earth's Civilizational Development

The study of alien civilizations necessitates a comparative examination of the development of human civilization on Earth. By drawing parallels and contrasts between the sociocultural, technological, and historical trajectories of human societies and hypothesized extraterrestrial civilizations, we can gain profound insights into the potential diversity of intelligent life in the cosmos. The process of comparative analysis involves identifying key milestones in terrestrial civilization, such as the advent of written language, the development of complex social structures, technological advancements, and the impact of cultural and ideological paradigms on societal evolution. While Earth's civilizations have followed a linear path of growth, marked by distinct periods of innovation and transformation, it is essential to acknowledge that any potential comparison with alien civilizations may be inherently limited by our terrestrial biases and perspectives. However, this limitation should not discourage us from engaging in this critical exercise. The comparative analysis of Earth's civilizational development also extends to the exploration of ethical, moral, and philosophical dimensions. As we contemplate the potential existence of advanced extraterrestrial societies, we are compelled to consider the ethical paradigms that have shaped human interactions and governance. Through a comparative lens, we can examine how differing cultural, religious, and ideological frameworks have impacted the trajectory of human history, inviting reflection on how these factors may manifest in hypothetical alien civilizations. Furthermore, the comparative analysis encompasses a comprehensive consideration of technological progress and its implications for societal evolution. By examining the parallels and disparities between Earth's technological milestones and speculated extraterrestrial advancements, we can broaden our understanding of the possible trajectories of intelligent beings elsewhere in the universe. Additionally, this analysis catalyzes contemplating the cosmic implications of technological progress, including the potential interstellar dissemination of knowledge, resources, and ideologies. Although speculative, the comparative analysis of Earth's civilizational development offers an invaluable framework for the exploration of cosmic diversity and the delineation of potential commonalities and distinctions between humanity and any putative alien civilizations. This intellectual endeavor enriches our capacity to perceive the vast tapestry of intelligent life in the cosmos and engenders a holistic understanding of our place within the broader narrative of cosmic existence.

Challenges in Research and Validation

As we venture into the study of alien civilizations, we are confronted with a myriad of challenges in research and validation. The foremost challenge lies in obtaining credible and verifiable data pertaining to extraterrestrial societies. Unlike terrestrial civilizations, which leave behind tangible artifacts and records, extraterrestrial civilizations may communicate and express their existence through incomprehensible means. Deciphering these signals and understanding their significance pose a significant hurdle in validating the existence of alien societies. Additionally, the lack of direct interaction or observation presents a barrier in corroborating our findings. The inability to visit or engage with these distant civilizations limits our ability to validate our hypotheses and theories. Furthermore, the vast distances between celestial bodies and the limitations of our current technology impede direct observation and validation of these civilizations. Another substantial challenge is the potential biases and preconceptions that may influence our interpretations and analyses. Our human-centric perspectives and cultural lens can inadvertently shape our understanding of alien civilizations, leading to subjective interpretations and conclusions. Moreover, the inherent complexity and diversity of potential extraterrestrial cultures further complicate the validation process. Each civilization may possess unique socio-cultural structures, languages, and technological advancements, making it arduous to generalize our findings across different civilizations. The lack of standardized methodologies and frameworks for studying alien civilizations also presents a formidable challenge. Traditional archaeological and anthropological tools developed for Earth-based studies may not be directly applicable to extraterrestrial contexts. As such, a concerted effort to adapt and innovate new methods and approaches tailored to cosmic archaeology is imperative. Additionally, the interdisciplinary nature of alien civilization studies necessitates collaboration among various scientific disciplines, including astronomy, linguistics, sociology, and philosophy. Achieving consensus and coherence across these diverse fields presents its own set of challenges. Ensuring the rigor and reliability of our findings in the absence of empirical evidence demands a high standard of scrutiny and validation. The hypothetical and speculative nature of our inquiries requires a balance between imaginative exploration and empirical substantiation. Overcoming these challenges will be instrumental in advancing our understanding of alien civilizations and shaping the future of cosmic archaeology.

Future Prospects in the Study of Alien Civilizations

As we look ahead to the future, the study of alien civilizations presents both exciting possibilities and daunting challenges. One of the key areas for future exploration revolves around the advancement of technology for detecting and analyzing potential extraterrestrial signals or artifacts. With rapid advancements in space telescopes, spectroscopy, and data processing techniques, researchers are poised to delve deeper into the cosmos and gather more nuanced insights into potential alien civilizations.

Additionally, the field stands to benefit greatly from interdisciplinary collaboration. By merging the expertise of astronomers, astrophysicists, archaeologists, anthropologists, linguists, ethicists, and other specialists, new perspectives and methodologies can be brought to bear on the question of extraterrestrial life. The synergy of diverse disciplines will enhance our understanding and aptitude for deciphering possible alien communications, decoding cultural artifacts, and discerning any potential technological signatures left by non-terrestrial civilizations.

Furthermore, future research endeavors may turn toward investigating the prospect of interstellar diplomacy. The ethical and political implications of potentially encountering alien civilizations are profound. Discussions surrounding the best course of action, protocols for communication, and considerations of mutual respect and understanding will become increasingly pertinent. Engaging with these topics now can help lay the groundwork for a harmonious approach to any future encounters.

The development of advanced artificial intelligence and machine learning may also revolutionize the field. Autonomous systems capable of sifting through vast datasets and identifying potential patterns or anomalies could prove invaluable in the search for evidence of alien civilizations. These technological advances have the potential to both streamline and enhance our efforts in uncovering traces of extraterrestrial existence.

Moreover, looking forward, the study of alien civilizations must not neglect the possibility of fundamentally different forms of life. While much of our current focus is on seeking biosignatures similar to those found on Earth, considering the potential for alternative biochemistries and modes of existence expands the scope of our exploration. This opens up rich avenues for astrobiological research that may lead to profound discoveries beyond our current imagination.

In conclusion, the future of the study of alien civilizations holds great promise. Advancements in technology, interdisciplinary collaboration, ethical deliberations, and the embrace of new paradigms all stand to shape the trajectory of this captivating field. As we venture further into the cosmos, the quest to understand our place in the universe and the potential existence of other sentient beings remains one of the most exhilarating intellectual and philosophical frontiers of our time.

Chapter Thirteen

Astrobiological Evidences of Life Beyond Earth

Astrobiology, the interdisciplinary study of the origin, evolution, and distribution of life in the universe, represents a fundamental quest to comprehend the existence of life beyond Earth. It is an endeavor that unites various scientific domains, such as astronomy, biology, chemistry, and geology, in an exploration of the potential for life in the cosmos. The implications of astrobiological research extend far beyond the bounds of terrestrial biology, offering profound insights into the nature of life on Earth and its place in the wider cosmic context. This emerging field seeks to address one of the most profound questions of human existence: are we alone in the universe? The search for extraterrestrial life has captivated the imagination of scientists, philosophers, and the public at large, driving a concerted effort to unravel the enigma of life's cosmic prevalence. Beyond mere curiosity, astrobiological research holds immense pragmatic value, with the potential to inform our understanding of planetary habitability, guide future space exploration endeavors, and perhaps even illuminate the origins of life itself. By delving into the conditions necessary for life to emerge and thrive within diverse astronomical environments, astrobiology enables us to contemplate the likelihood of encountering alien life forms and elucidate the circumstances under which they may exist. As humanity embarks on ambitious missions to explore Mars, Europa, Enceladus, and exoplanets beyond our solar system, the insights gained from astrobiological investigations stand to shape our perception of our place in the grand tapestry of the cosmos. Through the synergy of cutting-edge research methodologies, technological innovations, and theoretical paradigms, astrobiology presents a unique opportunity to unify scientific, philosophical, and cultural perspectives in pursuit of the ultimate question: what does it mean to be alive in the universe?

Chemical Markers of Life: Tracing Biological Compounds in the Cosmos

Detection of chemical markers of life in the cosmos is a key focus of astrobiological research. Various biological compounds play a crucial role in identifying potential extraterrestrial life. The search for these compounds involves scrutinizing diverse environments using advanced instrumentation and techniques. This pursuit requires a multidisciplinary approach, combining expertise from chemistry, biology, astronomy, and geology. Scientists seek to identify biomolecules such as amino acids, nucleic acids, lipids, and other organic molecules that are indicative of life as we know it. Beyond the familiar compounds found on Earth, researchers also explore the possibility of unconventional biochemistry on other planets or moons. The ability to trace these chemical markers of life in the cosmos requires innovative methodologies. Analytical techniques including mass spectrometry, nuclear magnetic resonance spectroscopy, and chromatography enable scientists to detect and characterize complex organic compounds. Additionally, spectroscopic analyses of planetary atmospheres and surfaces provide valuable insights into the presence of biogenic molecules. The search for chemical markers of life encompasses a wide range of celestial bodies, from exoplanets to moons within our own solar system. For instance, studies of Mars have revealed tantalizing evidence of organic molecules, suggesting the planet may have hosted habitable environments in the past. Similarly, investigations of the icy moons Europa and Enceladus indicate the presence of subsurface oceans that could harbor the building blocks of life. Cutting-edge instruments onboard space missions, such as the Mars rovers and upcoming missions to Jupiter's moons, are equipped to search for biosignatures and evaluate the potential habitability of these distant worlds. Considering the vastness of the cosmos, the identification of chemical markers of life beyond Earth is both a complex scientific endeavor and a profound existential quest. It holds the promise of unraveling the mysteries of our cosmic origins and addressing one of humanity's most profound inquiries: are we alone in the universe?

Planetary Habitats: Identifying Viable Ecosystems Beyond Earth

The search for viable ecosystems beyond Earth is a fundamental aspect of astrobiological research. When considering the potential for extraterrestrial life, it is crucial to identify planetary habitats that could support biological activity. Planetary habitats encompass a wide range of environmental conditions, from the surface features of a planet to its atmospheric composition and presence of water. Understanding these factors is essential in determining the potential habitability of exoplanets and celestial bodies within our solar system.

One key focus of this quest is the identification of exoplanets within the habitable zone of their host stars. This region, also known as the Goldilocks zone, offers conditions where liquid water could exist on the planet's surface. The presence of liquid water is a critical factor for the emergence and sustenance of life as we know it. Scientists leverage advanced telescopes and observational techniques to detect exoplanets within this region and analyze their atmospheres for signs of habitability.

Moreover, the exploration of our own solar system provides intriguing insights into potential planetary habitats. Mars, with its history of water flow and the presence of ice, has captivated researchers seeking evidence of past or present life. Europa, one of Jupiter's moons, has a subsurface ocean that raises tantalizing possibilities for microbial life in an entirely different environment. These investigations require a multifaceted approach, utilizing robotic missions, remote sensing tools, and sample return missions to unravel the mysteries of these distant worlds.

Furthermore, indigenous lifeforms on Earth have expanded our understanding of potential planetary habitats. Extremophiles - organisms capable of thriving in extreme conditions such as high temperatures, acidic environments, or high-pressure settings - serve as analogs for potential extraterrestrial life. By studying extremophiles, scientists can assess the adaptability of life under harsh conditions and extrapolate these findings to bolster our search for viable ecosystems beyond Earth.

As technology continues to evolve, the identification of viable planetary habitats becomes increasingly refined. Future advancements in space exploration, such as the deployment of next-generation telescopes and the development of sophisticated life-detection instruments, will enhance our capabilities to scrutinize exoplanets and celestial bodies with unprecedented detail. This relentless pursuit will bring us closer to answering the age-old question of whether life exists beyond our home planet.

Techniques in Detecting Bio-signatures

The search for extraterrestrial life beyond Earth relies heavily on the detection of bio-signatures - evidence of past or present life. This endeavor involves a complex array of techniques and methods designed to identify potential indicators of biological activity within diverse planetary environments. One prominent approach entails spectroscopy, which entails analyzing the interaction between matter and electromagnetic radiation. Spectral analysis of exoplanetary atmospheres enables scientists to detect specific molecules that may hint at the presence of life-sustaining processes. Additionally, the study of isotopic ratios plays a crucial role in distinguishing between the biotic and abiotic origins of certain compounds. By observing the relative abundance of stable isotopes such as carbon, hydrogen, nitrogen, and oxygen, researchers can infer valuable insights about the history of potential alien biospheres. Furthermore, the use of biomarkers—distinctive organic compounds indicative of past or present life—remains integral to bio-signature detection. These may include specific amino acids, lipids, or nucleic acids that are characteristic of terrestrial organisms, albeit potentially adapted and modified in an extraterrestrial environment. In tandem with spectroscopy, the technique of high-resolution imaging offers a means to potentially visualize bio-signatures in the form of microbial mats, stromatolites, or other sedimentary structures that could be indicative of ancient life forms. To complement these methods, ongoing research in the field of paleontology serves as a valuable reference for understanding terrestrial life's fossil record and how this knowledge can inform the search for similar evidence beyond our planet. As technology and scientific knowledge continue to advance, cutting-edge instruments such as mass spectrometers, chromatographs, and ultra-sensitive detectors are becoming increasingly vital in identifying minute traces of potential bio-signatures in extraterrestrial environments. The combination of these approaches represents a multidisciplinary effort to establish robust criteria for detecting life beyond Earth, serving as a testament to humanity's ceaseless curiosity and determination to unravel the cosmic enigma.

Interstellar Organic Molecules and Their Origins

Interstellar space, a vast expanse littered with molecular clouds, serves as the breeding ground for the genesis of organic molecules. These complex structures, comprising carbon-based atoms intricately bonded with hydrogen, oxygen, nitrogen, and other elements, originate through a series of intricate chemical processes occurring in the cosmos. The presence of these interstellar organic molecules, also known as astrochemical precursors, poses a profound implication for the potential existence of life beyond Earth. Understanding their origins and distribution within the cosmos is crucial in unraveling the cosmic enigma. One of the primary mechanisms through which interstellar organic molecules are formed is through the energy-rich environments of star-forming regions. Here, ultraviolet radiation, cosmic rays, and shock waves trigger chemical reactions within the interstellar medium, leading to the synthesis of diverse organic compounds. Furthermore, the remnants of stellar explosions, such as supernovae or novae, contribute to the enrichment of the interstellar medium with heavy elements essential for the formation of complex organic molecules. The study of these organic compounds using advanced spectroscopic techniques provides invaluable insights into the molecular composition of interstellar space, revealing the rich diversity of organic chemistry occurring outside our home planet. The identification of these interstellar organic molecules not only sheds light on the chemical processes shaping the cosmos but also offers compelling evidence for the cosmic distribution of the building blocks of life. Moreover, the presence of these molecules in protoplanetary disks and planetary systems underscores their potential role in seeding developing planets with the fundamental constituents necessary for life as we know it. As research in astrochemistry advances, the intricate pathways leading to the formation of interstellar organic molecules continue to captivate scientific inquiry, fueling a deeper understanding of the cosmic forces at play. The quest to comprehend the origins and distribution of these molecules in the vast celestial arena stands as a testament to humanity's enduring curiosity and determination to unravel the mysteries of the universe.

Role of Extremophiles in Understanding Alien Life

Extremophiles, as the name suggests, are organisms that thrive in extreme environments on Earth, such as high temperatures, acidic conditions, or high salinity. These resilient microorganisms have opened new avenues for understanding the potential for life beyond Earth. By studying extremophiles, scientists can gain insights into the adaptive strategies and biochemical pathways that could enable life to exist in extraterrestrial habitats with harsh conditions. The implications of these findings are profound, as they challenge traditional notions of habitable zones and broaden the scope of astrobiological exploration. Extremophiles also offer valuable clues for identifying potential biosignatures in the search for extraterrestrial life. For instance, extremophiles like thermophiles, which thrive in high-temperature environments, provide a framework for understanding how organisms might withstand extreme heat on planets or moons with volcanic activity. Similarly, halophiles, adapted to high-salinity environments, shed light on the potential for life in briny oceans on other celestial bodies. The study of extremophiles goes beyond Earth-centric perspectives and fosters a paradigm shift in astrobiological research by expanding the definition of habitability. Furthermore, extremophiles have practical applications in space exploration, as their unique biochemistry and resilience inform the design of instruments and experiments for detecting and studying alien life forms. This intersection of extremophile research and astrobiology presents an exciting frontier for interdisciplinary collaboration and innovation. As humanity ventures further into the cosmos, the role of extremophiles in shaping our understanding of alien life and habitable environments will undoubtedly continue to captivate and inspire scientists, philosophers, and enthusiasts alike.

Bioinformatics in Astrobiology: Computational Approaches

Bioinformatics plays a crucial role in astrobiology, serving as a vital bridge between biological data analysis and the exploration of potential extraterrestrial life. In the quest to understand the origins and potential existence of life beyond Earth, researchers have turned to computational approaches to glean insights from vast datasets and complex biological information. The field of bioinformatics provides the tools and methodologies necessary to process and analyze biological data, and its application in astrobiology holds immense promise for unraveling the mysteries of alien life.

At the heart of bioinformatics lies the integration of information technology, computer science, and biology. Computational algorithms are employed to mine massive repositories of genomic, proteomic, and metabolomic data, aiming to discern patterns, similarities, and evolutionary relationships that may shed light on the potential forms of life in other celestial bodies. Leveraging powerful computational techniques, astrobiologists can unravel the intricate web of genetic sequences and biological processes that could underpin extraterrestrial life.

One of the primary challenges in astrobiology is the identification of bio-signatures or indicators of past or present life on other planets. Bioinformatics offers a multidisciplinary approach to this task, employing computational modeling and statistical analyses to identify potential biomarkers within the spectral signatures of exoplanetary atmospheres. By harnessing sophisticated algorithms and data-driven models, researchers aim to distinguish biological features from abiotic processes, offering a systematic framework for evaluating the potential habitability of distant worlds.

Furthermore, bioinformatics enables astrobiologists to explore the concept of universal biology—a fundamental exploration of life's fundamental principles that transcend terrestrial constraints. By scrutinizing genomic sequences, molecular structures, and biochemical pathways through a computational lens, scientists seek to delineate plausible scenarios for alien life based on shared molecular frameworks and evolutionary mechanisms. This comparative approach not only informs our understanding of potential extraterrestrial biospheres but also enhances our perspective on the interconnectedness of life within the cosmos.

In light of the burgeoning field of exoplanet discoveries, bioinformatics stands as an indispensable toolkit for assessing planetary environments and their potential to harbor life. Through machine learning algorithms, statistical analyses, and data visualization techniques, astrobiologists can sift through astronomical data to pinpoint promising candidates for further exploration. By leveraging advanced computational methods, such as artificial neural networks and pattern recognition algorithms, researchers are poised to uncover novel insights into the habitability of exoplanets and the prospects for discovering life beyond Earth.

The convergence of bioinformatics and astrobiology heralds a new frontier in the search for extraterrestrial life. As computational power continues to advance and our understanding of biological systems expands, the prospect of decoding the cosmic tapestry of life becomes increasingly tangible. Through interdisciplinary collaboration and the fusion of computational expertise with biological inquiry, the realm of bioinformatics emerges as an essential compass guiding the exploration of life beyond our celestial shores.

Promising Exoplanets

In the search for extraterrestrial life, astronomers have identified several exoplanets that show potential for habitability and the presence of life-sustaining conditions. These promising exoplanets offer a glimpse into the diversity of planetary systems beyond our solar system and provide opportunities for further exploration and investigation. Among these promising candidates is Proxima Centauri b, an exoplanet located in the habitable zone of the nearest star to our solar system, Proxima Centauri. This rocky world has sparked immense interest due to its potential for hosting liquid water, a key ingredient for life as we know it. Additionally, TRAPPIST-1e, part of the TRAPPIST-1 planetary system, has emerged as an intriguing target for astrobiological studies. With its compact arrangement of seven Earth-sized planets, several of which orbit within the habitable zone, TRAPPIST-1e presents an enticing opportunity to explore potentially habitable environments beyond our solar neighborhood. Another compelling case is Kepler-452b, often referred to as "Earth's cousin" due to its similarities in size and orbital distance from its parent star. This exoplanet's sun-like star and location within the habitable zone make it a compelling prospect for studying the potential for life beyond Earth. Furthermore, LHS 1140 b has garnered attention for its rocky composition and proximity to its host star, making it a prime candidate for investigating the presence of an atmosphere and conditions conducive to life. The ongoing discovery and characterization of such promising exoplanets reflect the continuous advancements in observational techniques and instrumentation, underscoring the evolving landscape of astrobiology and the quest for understanding life beyond Earth. As researchers delve deeper into the exploration of these exoplanetary worlds, the data obtained from these case studies will inform and shape future missions, technology development, and theoretical models aimed at deciphering the enigma of life in the cosmos.

Implications for Future Missions and Research

As we explore the potential exoplanetary candidates that may harbor life, it becomes crucial to consider the implications for future missions and research endeavors in astrobiology. The identification of promising exoplanets through ongoing studies and observational data provides a basis for strategic planning and allocation of resources aimed at advancing our understanding of life beyond Earth.

Future missions will likely focus on the development and deployment of advanced space telescopes and observation platforms that are capable of conducting detailed analyses of exoplanetary atmospheres and surfaces. These instruments should be equipped with cutting-edge spectrometers and imaging devices to scrutinize the chemical composition, surface features, and potential biological activity in distant worlds. Additionally, the design and implementation of spacecraft and autonomous rovers tailored for interstellar exploration will be fundamental in enabling direct investigations of extraterrestrial environments.

In parallel, research efforts must emphasize interdisciplinary collaboration between astrophysicists, biologists, chemists, and planetary scientists to unravel the complexities of astrobiological systems. Integrated approaches involving computational modeling, laboratory simulations, and field expeditions will be necessary to contextualize the potential habitability of exoplanets and interpret the bio-signatures detected within their atmospheres.

The implications of future missions and research extend beyond scientific inquiry and technological advancement; they also pose ethical and philosophical considerations. As we penetrate deeper into the realm of exoplanetary exploration, questions surrounding the preservation of indigenous ecosystems, governance of extraterrestrial biological entities, and the impact of our discoveries on societal perceptions and beliefs arise. Addressing these concerns requires sustained dialogue among experts from diverse fields and a conscientious approach toward developing frameworks for responsible conduct in exploring alien environments.

Furthermore, the pursuit of future missions and research in astrobiology demands international cooperation and shared resources. Collaborative initiatives between space agencies, research institutions, and private enterprises will enhance the collective capability to undertake ambitious projects aimed at detecting and characterizing potential habitats for extraterrestrial life forms. Establishing global partnerships and information-sharing mechanisms will be instrumental in promoting transparency, fostering innovation, and mitigating challenges associated with technological, financial, and geopolitical barriers.

In conclusion, the implications for future missions and research in astrobiology are vast and multifaceted. They encompass not only scientific and technological dimensions but also ethical, societal, and collaborative aspects. By navigating these intricacies with foresight and integrity, humanity can embark upon a harmonized journey toward unraveling the enduring enigma of life beyond Earth.

Summary

The expedition into the realms of astrobiology is nothing short of a journey of wonder and discovery. As we embark on this quest to unravel the mysteries of life beyond Earth, it becomes increasingly evident that our terrestrial vantage point is just one small vantage point amidst an incomprehensibly vast cosmos. With each advancement in technology and methodology, we move closer to peering through the cosmic veil that shrouds potential extraterrestrial life forms from our understanding. The implications for future missions and research are manifold and wide-reaching. Our exploration of exoplanetary ecosystems and cosmic bio-signatures holds the promise of transforming our very understanding of life itself. Utilizing cutting-edge instruments and computational tools, astrobiologists are primed to detect key chemical markers indicative of life and habitable conditions beyond our planet. From the genetic makeup of extremophiles to the identification of interstellar organic molecules, every discovery paves the way for new endeavors. Equipped with these insights, future missions can be tailored to target celestial bodies with a higher likelihood of harboring life, guiding our search toward promising exoplanets teeming with potential biospheres. The interplay between the biological, geological, and astronomical sciences is pushing the boundaries of our knowledge, catalyzing interdisciplinary collaborations, and cultivating novel approaches. As we summarize our current strides in astrobiological research, we stand at the precipice of a truly transformative era. This chapter has set the stage for delving deeper into the technological signatures of ancient civilizations in space, where we will further contemplate the profound existential questions surrounding the existence of extraterrestrial life and its philosophical implications.

Chapter Fourteen

New Horizons

The Emerging Frontiers in Cosmic Exploration

The ongoing advancements in cosmic exploration have revolutionized our understanding of the universe and expanded the horizons of human achievement. As we continue to push the boundaries of what is currently possible, the emergence of cutting-edge technologies promises to propel space research into an unprecedented era of discovery and innovation. Nanotechnology, with its potential to manipulate matter at the atomic and molecular scale, is poised to revolutionize space exploration by offering solutions to challenges that have long hindered our endeavors beyond Earth's atmosphere. The integration of nanotechnology in space research presents a plethora of possibilities, from enhancing the durability of spacecraft materials to enabling the development of miniature yet powerful sensors for data collection. Moreover, the utilization of nanosatellites and nanorovers holds the promise of facilitating cost-effective and sustainable means of exploring celestial bodies, such as asteroids and moons, which were previously deemed inaccessible. This convergence of nanotechnology and space research underscores the potential for significant leaps in our ability to study and comprehend the cosmos. Looking ahead, foreseeing future developments in this field involves envisioning the deployment of nanobots for intricate tasks like repairing spacecraft while in orbit or constructing structures on extraterrestrial surfaces. Such advancements could mark a paradigm shift in cosmic exploration, allowing us to venture further into the depths of space and redefine the limits of what we can achieve. By pushing the frontiers of nanotechnology in conjunction with space research, scientists and engineers are set to pave the way for pioneering breakthroughs that will unlock the mysteries of the universe and shape the trajectory of humanity's cosmic journey.

Integration of Nano-Technology in Space Research

The integration of nano-technology in space research marks a paradigm shift in our approach to understanding the cosmos. Nano-technology is revolutionizing space research by offering unprecedented precision and functionality, enabling us to explore the complexities of space with greater depth and accuracy. One of the key applications of nano-technology in space research is the miniaturization of instruments, which allows for the development of smaller, yet highly efficient devices for various scientific purposes. These miniature devices, equipped with advanced sensors and communication systems, are capable of collecting and transmitting valuable data from remote cosmic locations, thus expanding our ability to explore and understand the vastness of space. Furthermore, nano-materials are being harnessed to enhance the resilience and durability of spacecraft, ensuring their ability to withstand the harsh conditions of space travel. By integrating nano-materials into spacecraft design, we can achieve lighter and stronger spacecraft structures, reducing mission costs and increasing payload capacity. Another crucial aspect of nano-technology in space research is the development of nano-satellites, also known as CubeSats. These small, low-cost satellites have opened up new avenues for space exploration, allowing for more frequent and affordable scientific missions. Their compact size and modular nature make them ideal for deploying a network of interconnected satellites, enabling collaborative and synchronized observations of celestial phenomena. Moreover, nano-technology facilitates the creation of innovative propulsion systems, such as nano-ion thrusters, which offer improved fuel efficiency and extended operational lifespans, essential for long-duration missions in deep space. The utilization of nano-technology in space research not only enhances our scientific capabilities but also presents opportunities for groundbreaking discoveries and technological advancements. As we continue to harness the potential of nano-technology, the boundaries of what we can achieve in space research are constantly expanding, paving the way for transformative breakthroughs that will shape the future of cosmic exploration.

Advancements in Cosmic Observation Instruments

The field of cosmic observation instruments has witnessed remarkable advancements in recent years, revolutionizing our ability to perceive and comprehend the cosmos. Innovations in telescope technology have enabled researchers to explore celestial bodies with unprecedented precision, leading to groundbreaking discoveries and a more profound understanding of the universe. One notable advancement is the development of adaptive optics, which mitigates the distorting effects of the Earth's atmosphere, allowing telescopes to capture clearer images of distant stars, galaxies, and exoplanets. Furthermore, the integration of advanced spectrographs has enhanced our capacity to analyze the chemical compositions of celestial objects. These breakthroughs have not only expanded our knowledge of astrophysics but also contributed to the search for extraterrestrial life.

In addition to ground-based observatories, space-based telescopes have played a pivotal role in augmenting our cosmic observation capabilities. Notably, the Hubble Space Telescope has provided unparalleled insights into the cosmos, capturing awe-inspiring images and collecting data that have significantly influenced our understanding of the universe's origins, evolution, and structure. Moreover, the upcoming James Webb Space

Telescope is poised to be a game-changer, with its advanced infrared capabilities promising to unveil new facets of cosmic phenomena that were previously beyond the reach of observational technologies. These developments highlight the instrumental role of space-based observation instruments in pushing the boundaries of astronomical exploration and reshaping our perceptions of the cosmos.

Furthermore, the fusion of traditional observational techniques with cutting-edge technologies such as machine learning and artificial intelligence has amplified the efficiency and precision of cosmic observation. The application of AI algorithms in processing vast volumes of astronomical data has facilitated the identification of subtle patterns and anomalies, accelerating the pace of discovery in the realm of astrophysics. Moreover, the utilization of advanced image-processing techniques has allowed astronomers to discern faint and distant cosmic objects, opening new avenues for exploration and inquiry. These advancements underscore the synergistic potential of integrating state-of-the-art technologies with established observational methods, propelling the frontiers of cosmic observation into uncharted territories.

As the landscape of cosmic observation continues to evolve, researchers and institutions must collaborate in nurturing the growth and dissemination of these technological marvels. The continued investment in research and development initiatives aimed at enhancing observation instruments will undoubtedly yield profound dividends, expanding our cognitive horizons and advancing our collective pursuit of unlocking the enigmas of the universe.

Quantum Computing: A Game Changer in Space Explorations

Space exploration has always been intricately tied to technological advancements that push the boundaries of what is achievable. As humanity continues to seek opportunities for interstellar travel and deeper cosmic understanding, the role of quantum computing emerges as a pivotal game changer in our quest for space exploration. Quantum computing fundamentally operates on the principles of quantum mechanics, enabling it to process vast amounts of data and perform complex calculations at exponentially greater speeds than traditional computers. In the context of space exploration, this capability holds transformative potential. One of the key applications of quantum computing in space exploration lies in its ability to optimize trajectory planning for spacecraft. The intricate calculations required for plotting efficient courses in the complex gravitational fields of celestial bodies can be expedited and refined through quantum algorithms, allowing for more precise and resource-efficient space missions. Additionally, quantum computing has the potential to revolutionize data analysis from deep space observations, empowering scientists to rapidly decipher and interpret massive volumes of cosmic data with unprecedented speed and accuracy. This could lead to groundbreaking discoveries and a deeper understanding of celestial phenomena. Furthermore, quantum computing presents opportunities for enhancing cybersecurity in space missions, safeguarding critical communication and navigation systems from potential threats. The development of secure quantum communication protocols could prove vital for maintaining the integrity and confidentiality of sensitive information exchanged over vast distances in space. Another area where quantum computing promises to make significant strides is in simulating complex phenomena relevant to space exploration, such as quantum simulations that model molecular behavior or simulate the properties of materials under extreme conditions, enabling the advancement of spacecraft design and material science. However, while the potential benefits of quantum computing in space exploration are immense, challenges remain. Overcoming the technical hurdles of scaling up and stabilizing quantum systems to operate reliably in the harsh conditions of outer space represents a significant obstacle. Moreover, ensuring the compatibility of quantum technologies with existing space infrastructure and protocols demands diligent attention. Nonetheless, as research and development in quantum computing continue to progress, the prospect of integrating this cutting-edge technology into space exploration endeavors holds tremendous promise for unlocking new frontiers and realizing ambitious cosmic aspirations.

Interstellar Travel: Theories and Potential Technologies

Interstellar travel, the ambitious pursuit of traversing the vast distances between stars, has long been a subject of fascination and scientific inquiry. In this section, we delve into the theoretical underpinnings and potential technologies that may someday enable humanity to embark on interstellar journeys.

The concept of interstellar travel involves overcoming immense challenges related to distance, time, energy, and the harsh conditions of outer space. Key theoretical frameworks such as warp drives, wormholes, and antimatter propulsion have captured the imagination of scientists and science fiction enthusiasts alike. While some of these concepts remain speculative or firmly entrenched in theoretical physics, they stimulate important discussions and inspire innovative research.

Recent advancements in propulsion technologies, including ion drives, solar sails, and nuclear thermal propulsion, have laid the foundation for more realistic approaches to interstellar travel. These emerging technologies promise greater efficiency, enabling spacecraft to achieve higher velocities and traverse longer distances within reasonable timeframes. Moreover, the development of advanced propulsion systems aligns with ongoing efforts to explore and exploit resources beyond our solar system.

Furthermore, the field of interstellar travel intersects with broader discussions on the potential colonization of exoplanets and the establishment of human outposts beyond our own stellar neighborhood. Ethical and existential considerations surrounding these endeavors prompt robust debates and necessitate careful consideration of the long-term implications of interstellar expansion.

As we look to the future, interdisciplinary collaboration among physicists, aerospace engineers, materials scientists, and astrophysicists will be pivotal in propelling the field of interstellar travel forward. Innovative research initiatives, such as Project Starshot and Breakthrough Starshot, showcase the dedication of scientists and entrepreneurs to push the boundaries of cosmic exploration and chart a course toward interstellar destinations.

In conclusion, while the realization of practical interstellar travel remains a distant ambition, the convergence of theoretical insights and practical advancements engenders optimism for the future. From theoretical models rooted in fundamental physics to experimental prototypes that challenge the boundaries of possibility, interstellar travel stands as a testament to humanity's enduring curiosity and quest for discovery.

Collaborative International Space Projects: Current and Future Scopes

In the rapidly evolving landscape of space exploration, collaborative international projects have emerged as crucial endeavors that leverage the collective expertise, resources, and technological capabilities of nations. By working together on ambitious space missions, countries can pool their strengths and address the complex challenges associated with the exploration of outer space. Current collaborative initiatives, such as the International Space Station (ISS), exemplify the power of joint ventures in pushing the boundaries of scientific knowledge and fostering global cooperation. These projects not only facilitate groundbreaking research across various scientific disciplines but also promote diplomatic ties and goodwill among participating nations. Looking ahead, the future of collaborative international space projects holds immense potential for expanding our understanding of the cosmos and driving technological innovation. As advancements in space technology continue to accelerate, international collaborations are expected to play an increasingly pivotal role in enabling groundbreaking discoveries and facilitating sustained human presence in space. Furthermore, new frameworks for cooperation are likely to emerge, paving the way for shared missions to distant celestial bodies, deep-space exploration, and long-term human habitation beyond Earth. The establishment of international partnerships will be fundamental in tackling the multifaceted challenges associated with extended space missions, including life support systems, astronaut health, and planetary protection. Moreover, as private-sector involvement in space activities grows, collaborative international projects can serve as models for public-private partnerships, propelling commercial space ventures while upholding high standards of safety, environmental responsibility, and ethical conduct. Central to the success of these projects is the spirit of mutual respect, transparency, and equitable resource sharing among participating nations. To harness the full potential of collaborative international space projects, it is imperative to foster open communication, information exchange, and knowledge transfer, aiming to cultivate a rich tapestry of diverse perspectives and expertise in the pursuit of common goals. By embracing the ethos of collaboration, the global community can embark on a new era of space exploration marked by unity, innovation, and collective achievement, laying the foundation for a future where humanity's reach extends far beyond the confines of our home planet.

Artificial Intelligence & Robotics: Enhancing Space Exploration Efficiency

The integration of artificial intelligence (AI) and robotics has revolutionized the landscape of space exploration, significantly enhancing efficiency and expanding the frontiers of our cosmic endeavors. AI technologies have enabled autonomous decision-making in space missions, leading to a reduction in human intervention and, consequently, minimizing the margin for error. From autonomous navigation to real-time data analysis, AI has proven instrumental in streamlining operations and optimizing resource management, thereby paving the way for more ambitious and extended missions beyond Earth's orbit. Furthermore, the deployment of robotics in space exploration has not only augmented the capabilities of human astronauts but has also facilitated the execution of complex tasks in harsh extraterrestrial environments. Robotic arms, rovers, and drones have been pivotal in conducting scientific experiments, collecting samples, and performing maintenance activities on space stations and planetary surfaces. These advancements have not only elevated scientific inquiry but have also laid the groundwork for future human colonization and industrialization of other celestial bodies. The synergy between AI and robotics has given rise to the development of sophisticated autonomous systems capable of adaptation and learning, further broadening the scope and potential of space exploration. Moreover, the fusion of AI with advanced robotics holds promising prospects for establishing sustainable habitats and infrastructure in outer space, thereby marking a pivotal juncture in humanity's quest to expand beyond the confines of Earth.

Addressing Ethical Considerations in Interplanetary Expansion

As humanity takes its first steps towards interplanetary expansion, a critical aspect that requires meticulous deliberation is the ethical considerations surrounding this endeavor. The pursuit of space exploration and potential colonization of other planets raises profound moral and ethical questions that command careful consideration. One of the primary ethical considerations revolves around the preservation and protection of extraterrestrial environments. As we venture beyond Earth, it becomes imperative to ensure that our activities do not disrupt or irreversibly damage the delicate ecosystems of other celestial bodies. This calls for stringent regulations and protocols to safeguard against contamination and ecological disruption on a planetary scale. Additionally, there is a pressing need to address the implications of human interference on indigenous extraterrestrial life forms, if they exist, and to establish guidelines for respectful coexistence. Ethical frameworks must be established to govern interactions with any discovered alien life, ensuring that scientific pursuits do not infringe upon the rights and welfare of such life forms. Another significant ethical consideration pertains to the equitable distribution of resources and opportunities in space exploration and colonization. As nations and private entities embark on ventures beyond Earth, it is vital to prevent exploitation and promote fair access to the benefits of cosmic exploration. This encompasses considerations of resource allocation, labor rights, and the prevention of monopolistic control over extraterrestrial territories. Moreover, the social and psychological well-being of individuals involved in interplanetary missions should be prioritized, encompassing concerns such as isolation, mental health, and the impact of prolonged space travel on interpersonal dynamics. Furthermore, ethical frameworks must address the potential consequences of cultural imposition and the preservation of diverse cultural identities in off-world settlements. Collaborative efforts among international institutions, policymakers, ethicists, scientists, and various stakeholders are paramount in developing comprehensive ethical guidelines for interplanetary expansion. These frameworks should align with timeless ethical principles while incorporating innovative approaches tailored to the unique challenges presented by space exploration. By proactively addressing these ethical considerations, humanity can aspire to build a sustainable and morally conscious foundation for our cosmic endeavors, ensuring that our advancements in the cosmos are characterized by integrity, responsibility, and respect for the broader cosmic community.

Economic Viability and Sustainability in Space Ventures

The exploration of space presents a myriad of opportunities for advancing scientific knowledge, technological development, and human understanding. However, the pursuit of space ventures must also be underpinned by economic viability and sustainability to ensure longevity and success in this frontier. This section examines the key aspects of economic viability and sustainability in space ventures, addressing the challenges and opportunities inherent in this domain.

One of the fundamental considerations in space ventures is the cost involved. Launching spacecraft, conducting research missions, and establishing extraterrestrial habitats require substantial financial investment. Therefore, evaluating the economic viability of such endeavors is essential. While government funding has traditionally been a significant driver of space exploration, there is a growing emphasis on commercialization and private sector involvement to diversify funding sources and sustain long-term initiatives. Private companies are increasingly contributing to space ventures, from satellite launches to asteroid mining, signaling a shift towards a more economically sustainable space industry.

In addition to economic viability, sustainability is a critical factor in space ventures. Sustainability encompasses not only the ecological impact of space activities but also their long-term feasibility and resilience. Space ventures must consider the environmental consequences of their operations, such as space debris management and resource utilization, to ensure responsible and sustainable practices. Moreover, maintaining economic sustainability over extended periods is crucial for the continuity of space exploration efforts, requiring careful planning, resource management, and innovation in business models.

Furthermore, economic viability and sustainability in space ventures are closely linked to international cooperation and collaboration. Shared infrastructure, joint missions, and collective investment in space projects can enhance economic efficiency, reduce costs, and facilitate mutual benefits. International partnerships promote the pooling of resources, expertise, and technology, fostering a more sustainable and economically viable approach to space exploration.

The commercial potential of space ventures extends beyond scientific research and exploration, encompassing satellite communications, tourism, manufacturing, and energy production. Establishing a robust space economy requires strategic planning, market analysis, and investment in technological advancements that can unlock new economic frontiers in space. Furthermore, considering the potential for space resources, such as rare minerals, water, and energy sources, presents immense opportunities for sustainable economic growth and development.

In summary, achieving economic viability and sustainability in space ventures necessitates a multi-faceted approach that integrates financial considerations, environmental responsibility, and international collaboration. Balancing economic interests with sustainable practices will be fundamental in shaping the future of space exploration and ensuring its enduring success.

Preparing for the Unexpected: Risk Management in Space Missions

Space missions, by their very nature, are complex and fraught with potential risks. As humanity ventures further into the cosmos, the need for robust risk management strategies becomes increasingly imperative. This section explores the multifaceted domain of risk management in space missions, encompassing proactive planning, comprehensive assessments, and adaptive responses to unforeseen circumstances.

The first critical aspect of risk management is identification and assessment. Space agencies and mission planners conduct thorough analyses to identify potential risks across all phases of a mission, from launch to landing. These risks range from mechanical failures and cosmic radiation exposure to supply shortages and communication disruptions. Utilizing extensive data and simulations, experts quantify the likelihood and potential impact of each risk, allowing for prioritization and resource allocation.

In tandem with identification comes mitigation and contingency planning. Mitigation involves implementing measures to reduce the likelihood or severity of identified risks. For instance, redundant systems, rigorous testing, and fail-safe mechanisms are integrated into spacecraft designs to minimize the impact of technical malfunctions. Concurrently, contingency plans are devised to address unforeseen occurrences. From emergency procedures for life support systems to alternative navigation strategies, these plans serve as crucial safeguards against potential crises.

Furthermore, effective risk management entails constant monitoring and adaptability. Throughout a mission, real-time data streams enable continuous assessment of potential risk indicators, allowing for proactive decision-making. Additionally, simulation models and scenario-based exercises are utilized to anticipate and prepare for various contingencies. The ability to swiftly modify mission parameters or initiate emergency protocols in response to unexpected events can be pivotal in averting catastrophic consequences.

Beyond technical considerations, human factors and psychological well-being are integral aspects of risk management. Understanding the psychological stresses and emotional resilience of astronauts is vital for maintaining mission functionality in high-pressure situations. Overcoming isolation, managing conflicts, and fostering strong team dynamics are essential components of ensuring psychological readiness for the unexpected.

In conclusion, navigating the unknown terrains of space demands an unwavering commitment to risk management. By embracing a comprehensive approach that encompasses meticulous assessment, strategic mitigation, and adaptive responsiveness, space missions can strive towards minimizing uncertainties and maximizing safety and success. Through relentless dedication to risk management, humanity can conquer the uncharted frontiers of space, pushing the boundaries of exploration and discovery.

Chapter Fifteen

Technological Signatures of Ancient Civilizations in Space

The exploration of technological artifacts left by ancient civilizations in space encompasses a fascinating intersection of archaeology, astronomy, and speculation about extraterrestrial intelligence. This emergent field seeks to probe the possibility that prior advanced life forms could have left tangible evidences of their existence beyond the confines of their native celestial bodies. At its core, the study of technological signifiers raises profound questions about the potential longevity and reach of intelligent beings and their footprints in the cosmic expanse.

In delving into the concept of technological signifiers, it is essential to first unravel the theoretical underpinnings that motivate such inquiries. The overarching framework involves contemplating the hypothetical trajectories of civilizations capable of interstellar travel or communication. This requires an examination of the sustainability of societies, the emergence of advanced technologies, and the possible manifestations and dispersal of these technologies across the cosmos.

Moreover, the concept of technological signifiers sparks contemplations about the temporal scales at which such artifacts might endure, the potential celestial destinations they might traverse, and the means by which they could be identifiable to future observers. With so many uncertainties surrounding the existence and characteristics of these artifacts, the chapter aims to navigate through the complexities and intricacies that underlie this futuristic endeavor.

Historical Context and Theoretical Framework

The search for technological signatures of ancient civilizations in space represents a critical endeavor that requires a thorough understanding of the historical context and a well-defined theoretical framework. Over centuries, human societies have pondered the possibility of extraterrestrial life and advanced civilizations in the cosmos. The historical context of this inquiry can be traced back to ancient myths, early philosophical discourse, and the evolution of scientific inquiry. Notable figures throughout history, including Galileo Galilei, Johannes Kepler, and Isaac Newton, speculated about the existence of otherworldly beings and their potential technological advancements.

In modern times, the development of radio astronomy, space exploration, and the search for exoplanets has intensified the pursuit of ancient technological artifacts beyond Earth. Theoretical frameworks for identifying and interpreting potential technological signatures must integrate diverse fields such as astrophysics, archaeology, anthropology, and astrobiology. This multidisciplinary approach allows researchers to consider a wide range of factors, including the potential for cross-cultural communication, the impact of interstellar distances on artifact dispersal, and the evolutionary trajectories of alien civilizations.

Moreover, the historical context and theoretical framework also encompass the socio-political dimensions of the search for ancient technological signatures in space. As humanity contemplates its place in the universe, questions of identity, cultural relativism, and the consequences of discovering extraterrestrial technologies become increasingly salient. These considerations are essential in framing the ethical and philosophical implications of encountering ancient technological artifacts, as they may challenge prevailing worldviews and societal structures.

Understanding the historical context and developing a robust theoretical framework is imperative for guiding research endeavors, shaping public discourse, and informing policy decisions related to the potential discovery of ancient extraterrestrial technologies. Moreover, it fosters an appreciation for the interconnectedness of human history, scientific exploration, and the ongoing quest to unravel the mysteries of the cosmos.

Identifying Technological Signatures: Methodologies and Challenges

As the field of Cosmic Archaeology evolves, one of the most compelling endeavors is the identification of technological signatures that could hint at the presence of ancient civilizations in space. This pursuit is beset with a multitude of methodological complexities and inherent challenges that necessitate a systematic and multidisciplinary approach. The first fundamental challenge lies in defining what constitutes a technological signature. At its core, a technological signature encompasses any discernible indication of advanced technological activity, ranging from anomalous electromagnetic emissions to artificial megastructures. However, differentiating such signatures from natural phenomena demands rigorous scientific scrutiny and meticulous data analysis. Methodologies for identifying technological artifacts revolve around leveraging cutting-edge instruments and novel observation techniques, often pushing the boundaries of contemporary scientific investigation. Remote sensing technologies play a pivotal role in this effort, facilitating the detection of potential techno-signatures across vast cosmic expanses. However, the interpretative ambiguity inherent in these observations necessitates complementary ground-based studies and validation processes using established benchmarks of technological activity on Earth. Additionally, the search for technological signatures encounters methodological impediments imposed by the vast temporal and spatial scales intrinsic to cosmic archaeology. Identifying artifacts of potential extraterrestrial origin requires confronting the enormity of cosmic distances and grappling with temporal incongruities, as well as accounting for the dynamism of cosmic phenomena. Furthermore, the search for technological signatures must navigate the enigma of cosmic context, acknowledging the need to differentiate between natural cosmic occurrences and purported technological anomalies. Consequently, conclusive identification demands a holistic assessment integrating archaeological, astrophysical, and astrobiological perspectives. Crucially, perhaps the paramount challenge confronting the recognition of technological signatures lies in avoiding interpretation biases and premature assumptions, thereby ensuring that scientific rigor underpins all stages of analysis. This underscores the imperative for interdisciplinary collaboration, uniting experts in diverse fields to scrutinize and validate potential techno-signatures using a balanced and dispassionate approach. Such collaborations are vital for fostering robust methodologies and mitigating the risk of false positives derived from narrowly focused investigations. Simultaneously, it is essential to acknowledge the profound philosophical and ethical dimensions implicated in ascribing cosmic techno-signatures to ancient civilizations, demanding conscientious consideration of the implications arising from such discoveries. Consequently, navigating the labyrinthine landscape of identifying technological signatures demands a pragmatic blend of scientific acumen, methodological innovation, and ethical perspicacity.

Case Studies of Suspected Technological Artifacts

Case Studies of Suspected Technological Artifacts The study of potential technological artifacts spread throughout the cosmos presents an exciting and complex challenge for cosmic archaeologists. By examining and analyzing diverse case studies, we gain insight into the potential existence of ancient civilizations in space and their associated technological signatures. One such case study involves the enigmatic structures on the moon's surface. Various anomalous formations, such as the intriguing presence of rectangular shapes and unusual patterns, have captured the attention of researchers, sparking debates about potential artificial origins. Analyzing these lunar anomalies introduces fascinating questions about the possibility of extraterrestrial technology existing within our own celestial neighborhood. Moving beyond our lunar neighbor, another compelling case study revolves around the peculiar objects discovered in orbit around distant stars. These objects display irregular motions and unexpected behaviors, prompting speculation about their artificial nature. Delving into the examination of these space-borne artifacts provides a glimpse into the complexities of identifying and interpreting potential technological relics in remote regions of the universe. Furthermore, the detection of unusual radio signals originating from deep space has spurred investigations into the potential existence of advanced cosmic communication systems. Instances of recurring and structured signals have raised tantalizing prospects of alien technological activities. Examining and documenting such enigmatic signals represents a crucial facet of the quest to unearth evidence of technologically adept extraterrestrial civilizations. Additionally, the Mars rover missions have yielded findings that prompt contemplation regarding the potential remnants of ancient technological activity on the Red Planet. Anomalies such as geometrical formations and metallic-looking objects have ignited discourse about the potential technological artifacts of a bygone era. Engaging with these Martian discoveries allows for a comprehensive examination of potential technological interventions in our neighboring planetary system. As we explore these case studies, we confront the intricate task of distinguishing between natural phenomena and potential technological relics. The rigorous application of scientific methodologies, technological advancements, and interdisciplinary collaboration plays a pivotal role in scrutinizing and understanding these suspected technological artifacts. By scrutinizing these captivating case studies, we gain valuable insights that guide our comprehensive approach to investigating potential extraterrestrial technological signatures and advance our understanding of the broader cosmic landscape.

Analytical Techniques in Artifact Examination

The examination of suspected technological artifacts from ancient civilizations in space requires a meticulous and multi-faceted approach. In combining traditional archaeological methods with cutting-edge scientific technologies, analysts must consider a diverse range of techniques to effectively study and interpret these enigmatic relics. One fundamental approach involves non-destructive testing methods such as X-ray fluorescence (XRF) spectroscopy, which can identify elemental composition without altering the artifact's integrity. Similarly, microscopic analysis plays a crucial role in scrutinizing minute details and surface characteristics, offering valuable insights into fabrication techniques and material origins. Additionally, scanning electron microscopy (SEM) enables researchers to explore the structural and chemical properties of artifacts at a microscale, shedding light on their manufacturing processes and potential extraterrestrial origins. The integration of advanced analytical techniques like mass spectrometry facilitates the detection of isotopic signatures, providing invaluable clues about the provenance and age of these artifacts. Furthermore, radiocarbon dating serves as a vital tool for establishing temporal context, allowing for the determination of the artifacts' antiquity. In conjunction with these scientific methodologies, interdisciplinary collaboration is essential, drawing expertise from fields such as materials science, astrophysics, and artificial intelligence. This holistic approach ensures a comprehensive and nuanced understanding of any suspected technological artifacts from ancient civilizations in space, offering profound implications for our comprehension of interstellar history and the technological prowess of advanced cosmic societies.

The Role of Artificial Intelligence in Decoding Tech Signatures

Artificial intelligence (AI) holds immense potential to revolutionize the field of interstellar archaeology through its capabilities to decode technological signatures left behind by ancient civilizations in space. Leveraging sophisticated algorithms and machine learning, AI offers a transformative approach to unraveling the mysteries encapsulated within enigmatic artifacts of extraterrestrial origin. By integrating AI-driven computational analyses with traditional archaeological methodologies, researchers can gain unprecedented insights into the nature, functionality, and purpose of these alien technological remnants. One of the primary advantages of employing AI in artifact decipherment lies in its capacity to discern patterns and correlations that may elude human cognition, thus facilitating the identification of intricate technological constructs embedded within seemingly ordinary objects. Furthermore, AI-driven image recognition and pattern detection algorithms enable the rapid detection of complex symbology or ideograms that might signify advanced scientific or engineering principles employed by ancient extraterrestrial civilizations. Collaborative efforts between AI specialists, archaeologists, and astrophysicists can harness the collective expertise to train AI models specifically for the recognition and interpretation of interstellar technological artifacts, paving the way for breakthroughs in understanding the cultural, scientific, and industrial achievements of alien societies. Moreover, AI systems can aid in reconstructing and simulating potential operational mechanisms of unearthed extraterrestrial devices, shedding light on their intended functions and practical applications. In addition to decoding singular artifacts, AI-based data analysis can facilitate the identification of latent connections or networks between disparate relics, unraveling broader technological landscapes indicative of advanced interstellar societies. As AI continues to evolve, the integration of natural language processing algorithms holds promise in deciphering inscriptions or textual fragments present on alien artifacts, potentially unveiling crucial knowledge encompassing domains such as astrophysics, cosmology, and the philosophical underpinnings of extraterrestrial cultures. While the application of AI in interstellar archaeology presents unparalleled opportunities for exploration and discovery, ethical considerations surrounding the preservation of cultural heritage and adherence to universal codes of conduct feature prominently. A responsible and conscientious approach is imperative to ensure that AI's analytical prowess is harnessed ethically and in alignment with the preservation of extraterrestrial heritage. As the frontiers of interstellar archaeology expand, the collaborative synergy between AI and human expertise promises to unlock the profound secrets concealed within the ancient technological vestiges scattered throughout the cosmos.

Interstellar Communication Systems: An Archaeological Perspective

The quest to unravel the mysteries of interstellar communication systems represents a pivotal facet of cosmic archaeology. Interstellar communication signals, if detected, could offer unprecedented insights into the technological prowess and cognitive capabilities of extraterrestrial civilizations. From an archaeological standpoint, the examination and interpretation of potential communication artifacts or transmissions pose a unique set of challenges and opportunities. At the forefront of this endeavor lies the need to discern between natural astrophysical phenomena and deliberate communication efforts. This distinction necessitates the deployment of cutting-edge analytical methodologies and interdisciplinary collaboration. Archaeologists, astrophysicists, computational linguists, and artificial intelligence experts must synergize their expertise to develop robust frameworks for identifying, analyzing, and understanding potential interstellar communication systems. The search for alien communication artifacts involves the sourcing, curation, and examination of anomalous signals or patterns within the cosmic microwave background, transient astronomical events, or targeted radio frequencies. Incorporating principles from information theory, cryptography, and linguistic analysis becomes indispensable in deciphering potential interstellar messages. Furthermore, the identification of potential carrier mechanisms for interstellar messages, such as neutrino beams or modulated laser emissions, demands meticulous consideration and examination. Envisioning a comprehensive approach to interstellar communication systems requires the comprehensive categorization and contextualization of any detected anomalous signals within the broader framework of Astro-archaeological investigations. Such a foundational lens empowers researchers to discern potential cultural, technological, or informational attributes embedded within the alleged communication artifacts. Moreover, the comparative analysis of terrestrial communication modalities, including ancient petroglyphs, pictograms, hieroglyphics, and contemporary digital data encoding, can furnish valuable reference points for contrasting and extrapolating potential extraterrestrial communication systems. The synthesis of knowledge pertaining to anthropological communication evolution alongside burgeoning studies in astrobiology and SETI augments the intellectual toolkit available for cognizing potential interstellar communication endeavors. Ultimately, by embracing an archaeological perspective on interstellar communication systems, the cosmic archaeologist becomes adept at gleaning profound insights into the societal structures, cognitive capacities, and expressive channels of extraterrestrial civilizations, thus catalyzing the transcendence of cosmic conjecture into tangible knowledge.

Comparative Analysis: Earth-based Technologies and Alien Counterparts

As we delve into the realm of extraterrestrial artifacts, it becomes essential to explore how these potential technologies compare to those developed on Earth. A comparative analysis between known human technologies and their hypothesized alien counterparts offers valuable insights into the technological evolution of civilizations across the cosmos. By examining the structural, functional, and material aspects of artifacts, we can identify parallels and divergences that shed light on the possible origins and purposes behind these enigmatic relics.

Earth's technological progress has been shaped by a confluence of cultural, environmental, and scientific factors. From ancient tools to modern electronic devices, our technological journey reflects a pattern of innovation driven by practical needs, societal dynamics, and scientific discoveries. This chronological progression serves as a framework for understanding the potential trajectories of technological development in extraterrestrial societies.

One intriguing avenue of comparison lies in the materials used in the construction of technological artifacts. While terrestrial technologies predominantly rely on naturally occurring elements and compounds, the speculated alien technologies may exhibit novel material compositions not found within our planetary boundaries. Advanced alloys, crystalline structures, or molecular configurations beyond terrestrial synthesis could mark a fundamental departure from familiar manufacturing processes, hinting at a level of scientific mastery exceeding our current capabilities.

Furthermore, the comparative study encompasses the functionality and purpose of technological artifacts. Human tools and devices are designed to serve specific functions, often varying according to cultural, economic, and technological contexts. Similarly, by discerning the intended functionalities of extraterrestrial artifacts, we gain insight into the potential societal needs, scientific ambitions, and cultural expressions of alien civilizations. The striking similarities or marked divergences can illuminate the underlying motivations driving technological innovation in distant corners of the cosmos.

This comparative analysis also extends to the structural and design features of artifacts, encompassing aspects of ergonomics, architecture, and engineering principles. Examining the forms and construction methods employed in alien technologies may unveil unconventional design paradigms, innovative engineering solutions, and aesthetic sensibilities that differ from terrestrial norms. Such insights offer a window into the cognitive processes, problem-solving strategies, and artistic inclinations of extraterrestrial engineers and artisans.

By juxtaposing Earth-based technologies with speculative extraterrestrial counterparts, we aim to construct a multidimensional framework that transcends cultural biases and terrestrial limitations. This comparative approach holds the potential to bridge the chasm between human experiences and cosmic possibilities, ultimately enriching our understanding of technology's place in the tapestry of universal civilizations.

Implications for Future Interstellar Archaeology

As we delve deeper into the realm of interstellar archaeology, contemplating the implications for our future pursuits becomes essential. The discovery and analysis of technological signatures from ancient civilizations in space lay the groundwork for a paradigm shift in human understanding and societal development. These implications transcend the boundaries of traditional archaeological investigation, propelling us into uncharted territories where imagination and science intertwine. One profound implication is the potential to decipher advanced extraterrestrial technology, which could revolutionize our own technological capabilities. By studying and understanding alien artifacts, we may gain insights into innovative approaches to energy generation, propulsion systems, and communication methods that surpass our current understanding. Furthermore, the identification of alien technological signatures holds the promise of fostering global collaboration and peaceful coexistence on an unprecedented scale. As we grapple with the realization that we are not alone in the cosmos, the societal impact of such discoveries cannot be overstated. The very fabric of human society and governance could undergo profound changes as we navigate the ethical, moral, and philosophical dilemmas associated with interstellar contact. Exploring and understanding technological artifacts from advanced civilizations beyond Earth also presents an opportunity to reevaluate our place in the universe and reassess our cosmic responsibilities. This redefinition of our cosmic identity could spur advancements in astrobiology, astronomy, and other scientific disciplines, reshaping the way we perceive our interconnectedness with the cosmos. Moreover, the pursuit of interstellar archaeology opens new frontiers for interdisciplinary collaboration, driving innovation across scientific, technological, and cultural domains. It beckons us to envision a future where the blurring of terrestrial and extraterrestrial knowledge systems leads to transformative breakthroughs that benefit humanity as a whole. Ultimately, the implications for future interstellar archaeology extend beyond scientific inquiry, transcending into the realms of culture, spirituality, and existential introspection. The journey towards comprehending the technological remnants of ancient civilizations in space marks a pivotal juncture in human history, igniting a collective quest for wisdom and understanding that stretches beyond the confines of our planet.

Summary

The quest for understanding ancient civilizations in space has led to a profound shift in our perspectives on existence, technology, and the universe. In the previous chapters, we have delved into the intricacies of cosmic archaeology, contemplated the implications of discovering technological signatures of ancient civilizations beyond Earth, and explored the potential pathways for future interstellar archaeological endeavors. As we conclude this exploration, it is imperative to reflect on the significance of these pursuits and chart a course for what lies ahead.

The compendium of knowledge amassed through the interdisciplinary approach of cosmic archaeology has unveiled the tantalizing possibility of ancient extraterrestrial civilizations and the remnants of their technological prowess scattered across the cosmos. This realization sparks a renaissance of thought, encouraging us to transcend the boundaries of terrestrial-centric ideologies and embrace the potential diversity of intelligent lifeforms in the universe.

Looking ahead, it becomes clear that the study of technological signatures of ancient civilizations in space presents both unprecedented opportunities and daunting challenges. The discovery of an artifact with undeniable extraterrestrial origins would revolutionize our comprehension of the cosmos, posing profound questions about the nature of intelligence, cultural evolution, and the interplay between different galactic societies. However, discerning the authenticity of such artifacts from natural phenomena or human-generated debris demands a rigorous scientific approach, advanced technological tools, and global collaboration among diverse experts.

Moreover, the ethical considerations inherent in interstellar archaeology cannot be understated. The impact of unearthing evidence of ancient extraterrestrial civilizations extends far beyond academic interest, touching upon philosophical, sociocultural, and geopolitical dimensions. As we envision the path forward, it is imperative to establish frameworks for responsible engagement with potential discoveries and develop guidelines that preserve the integrity of such findings while fostering international cooperation in navigating the societal implications of these revelations.

In navigating the path forward in interstellar archaeology, the integration of cutting-edge technologies and methodologies, including advancements in artificial intelligence, data analytics, astrobiology, and remote sensing, will be pivotal. These multidisciplinary approaches are poised to exponentially enhance our capacity for surveying, analyzing, and interpreting the traces of ancient technological achievements across cosmic landscapes, thereby propelling the field of cosmic archaeology towards new frontiers of exploration and understanding.

As we stand at this juncture of profound curiosity and unparalleled potential, it is essential to recognize that the quest for unraveling the enigmas of ancient civilizations in space extends far beyond individual achievements or institutional pursuits—it is an odyssey that transcends generations, encompassing the collective aspirations of humanity to unearth the mysteries of our cosmic ancestry.

Chapter Sixteen

Cosmic Ethics: Contemplating Moral Dilemmas Beyond Earth

As humanity sets its sights on the stars, we are confronted with a myriad of ethical quandaries that have never before been encountered. The dawn of interstellar exploration heralds a new chapter in our moral and philosophical evolution, demanding a reassessment of the fundamental principles that guide ethical decision-making. In the context of space exploration, the ethical landscape is boundlessly complex, as it requires us to contend with scenarios that lie far beyond the realms of terrestrial experience. The very act of venturing into the cosmos challenges us to reevaluate our understanding of autonomy, responsibility, and the preservation of life. At the heart of interstellar ethics lies the essential question of how we, as Earth's stewards, approach the unknown, how we communicate with extraterrestrial intelligence, and how we navigate the boundaries between discovery and intervention. It is within this framework that we must explore the foundation of ethical decision-making when dealing with the unknown aspects of space. Rooted in diverse philosophical perspectives and scientific insights, this examination prompts us to consider the implications of our actions not only for ourselves but also for potential alien civilizations and the broader cosmic ecosystem. As we delve into the intricate web of ethical considerations surrounding interstellar activity, we are compelled to engage in a reflective discourse about the values and principles that will guide our interaction with the enigmatic expanse beyond our home planet.

Ethical Considerations in Interstellar Communication

Interstellar communication presents a fascinating frontier for humanity, with the potential to engage with extraterrestrial civilizations. However, this endeavor also raises profound ethical considerations that necessitate careful examination. As we venture into the cosmic landscape of communication beyond Earth, we encounter intricate moral dilemmas that shape our interactions with unknown entities.

One of the foremost ethical considerations in interstellar communication revolves around the exchange of information and the potential impact on the recipient civilization. Transmitting knowledge to another civilization carries immense responsibilities, as it can influence their societal development, and worldview, and potentially alter the course of their history. This prompts crucial questions about the content, intent, and consequences of the messages we transmit. Ethicists and scholars emphasize the importance of considering the cultural, psychological, and social implications of our communications, urging thorough analysis before any transmission occurs.

Furthermore, the timing and procedures of initiating interstellar communication demand ethical scrutiny. The decision to make contact with extraterrestrial intelligences requires comprehensive assessment, encompassing deliberations about the potential consequences for both humanity and the contacted civilization. Delving into discussions surrounding the initiation of such communication involves weighing the potential benefits against risks, including unforeseen repercussions on global stability, cultural paradigms, and existential concerns.

In our exploration of ethical considerations in interstellar communication, we must also confront the issue of representing humanity as a whole. Crafting a message that accurately reflects human diversity, values, and aspirations while avoiding misinterpretation or offense becomes an intricate ethical challenge. Addressing this concern entails drawing from a broad spectrum of human experiences, perspectives, and languages to construct a coherent portrayal of our species, all within the bounds of respect, humility, and cultural understanding.

The ethical dimension of interstellar communication extends to the potential reception of extraterrestrial transmissions. Society must grapple with the implications of detecting and interpreting alien signals, contemplating the psychological, philosophical, and sociopolitical impacts that such revelations might evoke. It is imperative to establish frameworks for responsible analysis and disclosure of any received communications to navigate the intricate intersection of science, ethics, and public awareness.

In summary, ethical considerations in interstellar communication are multifaceted and pivotal, demanding deliberate reflection and conscientious decision-making. As we stand at the precipice of potential cosmic encounters, grappling with these profound ethical quandaries is essential in shaping a responsible and morally informed approach to engaging with the mysteries of the cosmos.

The Responsibility of Terraforming and Planetary Engineering

Terraforming and planetary engineering represent significant ethical challenges as humanity ventures beyond Earth to other celestial bodies. The act of terraforming involves transforming a planet, moon, or other body to make it more Earth-like and hospitable for human habitation. This process raises profound questions about our responsibility to the ecosystems and potential indigenous life forms that may exist on these extraterrestrial worlds.

One of the primary ethical considerations in terraforming is the preservation of existing environments and the possible disruption of indigenous ecosystems. It prompts a deep reflection on the value we place on the intrinsic worth of alien biospheres. The alteration of an entire planetary environment could lead to the irreversible extinction of unique life forms, and therefore, necessitates a thorough assessment of the moral implications involved.

Furthermore, the ethical dimension of terraforming extends to the unintended consequences of altering a planetary biosphere. Deliberate changes made to atmospheric composition, surface geology, and hydrology can have unpredictable effects on the overall ecology of the target world. When considering the potential impacts of such alterations, it becomes imperative to develop comprehensive risk assessments and mitigation strategies to minimize unforeseen harm to the planetary system as a whole.

Moreover, the notion of planetary engineering introduces the ethical quandary of imposing human will on extraterrestrial bodies. Is it ethically justifiable for humanity to assert dominion over entire worlds and sculpt them according to our preferences? This thought-provoking question delves into philosophical discussions surrounding human stewardship of the cosmos and the intersections of ethics, technological prowess, and cosmic expansion. Balancing the aspiration to explore and colonize space with a sense of humility and respect for the intrinsic qualities of other planetary realms is a fundamental issue at the heart of this ethical dilemma.

As our understanding of astrobiology advances, another ethical dimension emerges in the form of potential conflicts between terraforming efforts and the existence of indigenous extraterrestrial life. Contemplating scenarios where microbial, evolutionary, or intelligent life forms may already inhabit a target world necessitates careful contemplation of our role as intruders and the implications of our actions on the native biota. This calls for the establishment of rigorous protocols governing the conduct of our interactions with any existing extraterrestrial life, should it be discovered during the terraforming process.

In navigating the complex landscape of terraforming and planetary engineering, a meticulous incorporation of interdisciplinary perspectives from environmental ethics, biology, planetary science, anthropology, and law is indispensable. These diverse viewpoints can inform the development of ethical frameworks and governance systems that uphold responsible practices in our endeavors to transform extraterrestrial environments. Ultimately, the endeavor of terraforming and planetary engineering compels us to grapple with fundamental questions about our ethical obligations as stewards of the cosmos, calling for considered and informed decision-making to ensure harmonious coexistence with the broader cosmic community.

Protocols for Encounter: Interactions with Extraterrestrial Life

As humanity ventures further into the exploration of outer space, the prospect of encountering extraterrestrial life becomes an increasingly plausible scenario. Consequently, the development of protocols for interstellar encounters is an essential undertaking. This section delves into the intricate considerations and ethical frameworks that must be established in preparation for potential interactions with extraterrestrial life forms. The need for cautious and meticulous planning when crafting these protocols cannot be overstated. First and foremost, it is imperative to acknowledge the potential cultural, biological, and technological disparities between terrestrial beings and hypothetical extraterrestrial life forms. This acknowledgment serves as the foundation upon which the protocols for encounter are constructed. Moreover, the moral and ethical implications of initiating contact with alien civilizations underscore the urgency of delineating a comprehensive set of guidelines. Additionally, considering the vast spectrum of potential scenarios and communication methods, a multifaceted approach to protocol development is paramount. This comprises establishing strategies for communication, as well as implementing measures to mitigate potential risks and ensure the safety and preservation of both terrestrial and extraterrestrial life forms. Furthermore, engaging in interdisciplinary collaboration involving experts from fields such as anthropology, linguistics, biology, and ethics is instrumental in formulating robust and encompassing protocols. By synthesizing diverse perspectives, we can cultivate a holistic understanding of the multifaceted challenges associated with interstellar encounters. It is crucial to recognize that the establishment of these protocols extends beyond the confines of current terrestrial concerns and embraces a broader cosmic perspective. The guidance systems and foundational principles developed through the process of creating these protocols will inevitably shape the future of human interaction with the cosmos. Given the profound impact and far-reaching consequences of potential encounters with extraterrestrial life, conscientious and deliberate thought must underpin the formulation of these protocols. Ultimately, the development of protocols for encountering extraterrestrial life embodies our sense of responsibility and stewardship as cosmic citizens, serving as a testament to our commitment to ethically navigating the uncharted territories of the universe.

Conservation Ethics in Space Exploration

Conservation ethics play a pivotal role in shaping our approach to space exploration. As we venture into the cosmos and potentially encounter diverse ecosystems, it becomes imperative to establish ethical guidelines that prioritize the preservation of extraterrestrial environments and life forms. The exploration of space presents unique challenges as we navigate through uncharted territories, and the ethical considerations surrounding conservation become increasingly complex. It is essential for spacefaring nations and organizations to proactively address these concerns and develop frameworks that uphold the principles of environmental stewardship in cosmic endeavors. One of the fundamental aspects of conservation ethics in space exploration revolves around the concept of planetary protection. This involves implementing measures to prevent harmful contamination of celestial bodies and safeguarding them from potential terrestrial biological contaminants. By adhering to strict sterilization protocols for spacecraft and equipment, we can minimize the risk of unintentional microbial contamination on planets, moons, and other celestial bodies, thus preserving their indigenous environments. Furthermore, the ethical implications of terraforming and geoengineering efforts in space demand careful consideration. While these concepts hold the promise of transforming inhospitable worlds into habitable ones, they raise important questions regarding the long-term ecological balance and the potential disruption of indigenous ecosystems. Striking a harmonious balance between scientific exploration and conservation necessitates a comprehensive evaluation of the impact of human activities on extraterrestrial environments. Additionally, conservation ethics extend to the responsible utilization of extraterrestrial resources. As we contemplate the prospect of mining asteroids or harnessing resources from other celestial bodies, it is crucial to approach these endeavors with ethical foresight. Sustainable resource management practices, coupled with regulations that mitigate ecological harm, are essential in ensuring that our pursuit of cosmic resources does not compromise the integrity of alien habitats. In essence, the integration of conservation ethics into the fabric of space exploration endeavors underscores our commitment to being responsible custodians of the cosmos. By embracing a proactive approach to addressing environmental and ethical considerations in space exploration, we can aspire to create a future where our cosmic ventures coexist harmoniously with the ecological tapestries of the universe.

Ownership and Exploitation of Extraterrestrial Resources

The exploration and potential exploitation of extraterrestrial resources pose significant ethical considerations. As humanity reaches out into space, questions arise regarding the rightful ownership of celestial bodies and their resources. The legality and morality of claiming extraterrestrial territories become complex as we contemplate the implications of resource extraction beyond Earth. Various international agreements and legal frameworks have addressed aspects of space law, including the Outer Space Treaty of 1967, which prohibits any nation from claiming sovereignty over celestial bodies. However, the issue of resource exploitation remains a subject of ongoing debate and negotiation. One key consideration is the distinction between ownership and mere utilization of extraterrestrial resources. While the idea of owning celestial objects may conflict with the principle of preserving space as a common heritage of humankind, incentivizing responsible exploitation could encourage sustainable practices and further exploration. The ethical framework for extraterrestrial resource ownership must also incorporate principles of equity, ensuring that benefits are shared among all nations and not disproportionately acquired by a select few. Moreover, establishing mechanisms for oversight and regulation is imperative to prevent exploitation and ecological harm, resembling terrestrial conservation efforts. The potential formation of intergovernmental agreements or a global regulatory body could oversee the fair distribution and supervised use of extraterrestrial resources. Furthermore, transparency in resource utilization and environmental impact assessments will be crucial in gauging the ethical ramifications of extraterrestrial activities. Discussions of socially responsible mining and stewardship of extraterrestrial environments become integral components of considering the delicate balance between human advancement and cosmic preservation. As we look toward a future where space exploration and resource utilization are more tangible realities, the ethical approach to ownership and exploitation of extraterrestrial resources becomes paramount. It demands thoughtful reflection, international collaboration, and a shared commitment to upholding ethical standards that resonate with the values of humanity, even as we expand our presence beyond the confines of Earth.

Implications of Biotechnological Manipulations in Space

Biotechnology has been a driving force behind numerous scientific advancements on Earth, yet its potential applications and implications in space exploration raise complex ethical questions. The manipulation of biological organisms and processes to enhance human adaptation to extraterrestrial environments is an area of great significance and concern. One of the primary areas of consideration is the development of biotechnological interventions to mitigate the physiological challenges posed by extended space travel, such as muscle atrophy, bone density loss, and radiation exposure. While these interventions hold promise for safeguarding astronaut health, they also raise profound issues regarding medical ethics, consent, and the alteration of natural biological processes. Additionally, the prospect of genetically modifying organisms to thrive in extraterrestrial habitats necessitates a careful examination of the ecological impact and potential contamination of pristine environments. The ethical conundrums surrounding the creation, containment, and release of genetically engineered organisms beyond Earth's biosphere require comprehensive regulatory frameworks to prevent unintended consequences. Furthermore, the application of biotechnology in terraforming and planetary engineering introduces ethical dilemmas related to genetic manipulation, species preservation, and long-term ecosystem stability. As humanity expands its reach into outer space, the responsible and ethical use of biotechnological tools becomes increasingly paramount. This involves integrating principles of bioethics, environmental stewardship, and precautionary measures into the design and implementation of biotechnological initiatives in space exploration. Ultimately, grappling with the implications of biotechnological manipulations in space demands a conscientious and interdisciplinary approach, where ethical, scientific, and legal considerations are harmoniously woven together to guide humanity's endeavors beyond the confines of Earth.

Ethical Frameworks for Artificial Intelligence in Space Missions

As humanity continues to advance its exploration of outer space, the integration of artificial intelligence (AI) into space missions poses significant ethical considerations and challenges. AI systems are increasingly being employed to autonomously manage complex tasks such as spacecraft operations, data analysis, and decision-making processes in uncharted territories beyond Earth's atmosphere. The deployment of AI in these roles necessitates the establishment of robust ethical frameworks to govern their use and mitigate potential risks.

One critical aspect of delineating ethical frameworks for AI in space missions involves ensuring the accountability and transparency of AI decision-making processes. It is imperative to define clear lines of responsibility and oversight to prevent the misuse or unintended consequences of AI algorithms operating in isolated space environments. Establishing mechanisms for auditing and regulating the behavior of AI systems will be vital in maintaining trust and ethical integrity within the realm of autonomous space exploration.

Furthermore, the ethical deployment of AI in space missions demands a comprehensive assessment of its impact on potential extraterrestrial ecosystems and life forms. As we venture into uncharted cosmic domains, it is essential to consider the preservation of alien cultures and the mitigation of any detrimental effects that AI technology may impose on indigenous environments. Ethical frameworks must encompass protocols for assessing the ecological implications of AI-driven activities, including the identification and protection of vulnerable extraterrestrial habitats and species.

An overarching concern in the development of ethical frameworks for AI in space missions pertains to the safeguarding of human values and moral principles in the delegation of decision-making authority to AI systems. Deliberate efforts must be made to embed ethical guidelines and human-centric value systems within the programming and operational directives of AI technologies employed in space exploration. This includes prioritizing the avoidance of actions that could potentially compromise the dignity, autonomy, or rights of sentient beings encountered beyond our planet.

As the progression of space missions continues to intersect with rapid advancements in AI capabilities, international collaboration and consensus-building become paramount in the establishment of universal ethical standards for AI deployment in space exploration. Ethical frameworks should be informed by diverse perspectives and interdisciplinary expertise, fostering a collective commitment to responsible and sustainable practices in the evolving landscape of cosmic exploration. By proactively addressing the ethical implications of AI integration in space missions, humanity can uphold its moral obligations while embracing the transformative potential of AI technology in our quest to comprehend the cosmos.

Balancing Scientific Curiosity with Preservation of Alien Cultures

As humanity ventures further into the cosmos and encounters the possibility of extraterrestrial life, a critical ethical consideration arises: how do we balance our inherent scientific curiosity with the imperative to respect and preserve any encountered alien cultures? This dilemma is at the heart of space exploration and colonization, as the potential for coming into contact with intelligent extraterrestrial civilizations becomes an increasingly tangible prospect. Balancing the pursuit of knowledge with the preservation of alien cultures requires a sensitive and thoughtful approach, rooted in a deep appreciation for cultural diversity and the recognition of the intrinsic value of all sentient beings across the universe. One of the fundamental challenges in this endeavor lies in reconciling the inherently exploratory nature of scientific inquiry with the need to ensure that such exploration is conducted conscientiously and respectfully. This involves developing rigorous protocols for engagement with alien civilizations, guided by principles of mutual understanding, empathy, and peaceful coexistence. Moreover, it necessitates placing a strong emphasis on interdisciplinary collaboration, drawing insights from fields such as anthropology, sociology, and intercultural communication to inform the development of ethical frameworks for cross-cultural interactions in space. Additionally, embracing a mindset of cultural humility will be paramount in navigating these uncharted territories, acknowledging that our Earth-centric perspectives may not always align with the customs, values, and societal structures of extraterrestrial societies. Furthermore, the pivotal role of education and awareness cannot be understated, as fostering a public understanding of the significance of preserving alien cultures will be vital in shaping policies and decisions related to space exploration. This ethos of preservation encompasses not only the external behaviors and interactions of humanity with extraterrestrial civilizations but also extends to the internal dynamics of human society itself. It prompts introspection on the impact of our own cultural values and practices on our interactions with alien cultures, as well as the ways in which cultural heritage and diversity can enrich our collective understanding of the cosmos. In essence, the pursuit of scientific knowledge must go hand in hand with a profound reverence for the unique cultural tapestries that may exist beyond our planet, ensuring that our quest for discovery uplifts rather than undermines the existence of extraterrestrial civilizations.

Global Cooperation and Policy-making in Space Ethics

As humanity expands its presence in space and embarks on ambitious explorations beyond the confines of Earth, it becomes increasingly imperative to address the ethical considerations that accompany such endeavors. Global cooperation and policy-making in space ethics are essential components in ensuring that the exploration and potential exploitation of extraterrestrial environments is conducted responsibly and equitably. This section examines the intricate web of issues related to establishing international frameworks for space ethics and the complexities of formulating cohesive policies that harmonize diverse cultural, economic, and political perspectives.

At the heart of global cooperation in space ethics lies the recognition that the exploration and potential habitation of extraterrestrial environments transcends national boundaries. The establishment of comprehensive, universally accepted ethical guidelines necessitates collaboration among nations, international organizations, and other stakeholders. Key to this effort is the development of a framework that respects the principles of inclusivity, transparency, and accountability, thus fostering a collective sense of responsibility toward preserving the integrity of cosmic domains.

Crucial to the formulation of space ethics policies is the consideration of differing cultural perspectives and value systems. As humanity extends its reach into space, where contact with diverse life forms or the existence of unique ecosystems is plausible, the need to embrace cultural diversity becomes paramount. Discussions surrounding space ethics must therefore encompass the preservation of alien cultures, while simultaneously respecting the autonomy and heritage of indigenous civilizations that may be encountered in the vastness of the cosmos.

Moreover, addressing the socio-economic implications of space exploration and resource utilization within an ethical framework presents a multifaceted challenge in policy-making. Balancing the drive for innovation and progress with the ethical use of extraterrestrial resources requires careful deliberation and negotiation. Equity, fairness, and just distribution of benefits emerge as central tenets around which global cooperation in space ethics pivots.

Comprehensive policy-making also demands anticipatory measures to navigate emergent technologies and their potential impact on space ethics. Issues such as genetic engineering, artificial intelligence, and bioethics in space exploration call for a forward-looking approach that integrates scientific advancements with ethical considerations. International collaboration can facilitate the establishment of protocols that guide the ethical deployment and regulation of these technologies in extraterrestrial environments.

In conclusion, global cooperation and policy-making in space ethics embody an endeavor of profound significance as humankind steps into the cosmic arena. It prompts us to weave a tapestry of values, norms, and laws that uphold the dignity of all sentient entities and sustain the wondrous tapestry of the universe. Through collaborative efforts and thoughtful deliberation, humanity can carve a path that transcends national boundaries and fosters a shared commitment to ethically engaging with the cosmic realm, thereby paving the way for a future in which space exploration coexists harmoniously with the diverse life forms and cultures that may be encountered amidst the stars.

Chapter Seventeen

Future Frontiers: Expanding the Search for Cosmic Echoes

Advancements in Cosmic Exploration Technologies

In the realm of cosmic exploration, technological advancements are pivotal in unraveling the mysteries of the universe. The latest developments in telescopes, rovers, and satellites have redefined our capacity to peer deeper into space and uncover celestial phenomena that were once beyond our reach. Telescopes stand at the forefront of this technological revolution, with the inception of next-generation instruments equipped with cutting-edge optics and detectors that enhance resolution and sensitivity. The deployment of space-based telescopes, such as the James Webb Space Telescope, has opened new frontiers in astronomical observations by capturing unprecedented images of distant galaxies, exoplanets, and cosmic events. Furthermore, ground-based telescopes with adaptive optics and large apertures enable sharper imagery, enabling astronomers to study faint objects with remarkable detail. Complementing telescopic observations, planetary rovers represent another leap in cosmic exploration technologies. Advanced robotic explorers, like the Mars rovers, have expanded our understanding of other celestial bodies by conducting on-site analyses, collecting samples, and transmitting valuable data back to Earth. These rovers are equipped with a suite of scientific instruments, including spectrometers and cameras, allowing them to investigate the geological composition and environmental conditions of extraterrestrial landscapes. In addition to telescopes and rovers, satellites play a crucial role in widening the scope of cosmic research. Satellite missions designed for astrophysical studies provide a vantage point above Earth's atmosphere, enabling uninterrupted observations across different wavelengths of the electromagnetic spectrum. Moreover, space-borne observatories equipped with state-of-the-art instrumentation facilitate studies in areas such as cosmology, exoplanet detection, and high-energy astrophysics. Together, these advancements in cosmic exploration technologies pave the way for groundbreaking discoveries, fostering a deeper comprehension of the cosmos and the potential for unearthing clues about the origins and evolution of the universe.

Integration of AI in Cosmic Search Strategies

The integration of artificial intelligence (AI) in cosmic search strategies represents a transformative advancement in the field of space exploration. AI facilitates the automation of complex tasks, enabling the analysis of vast datasets at speeds far surpassing human capabilities. In the context of cosmic search strategies, AI plays a critical role in sifting through astronomical observations, identifying patterns, and discerning potential signals from cosmic phenomena. With the exponential growth of data generated by telescopes, satellites, and other space probes, AI has become indispensable in processing, categorizing, and interpreting this immense influx of information. Machine learning algorithms, a subset of AI, empower researchers to detect subtle anomalies that may signify previously unidentified cosmic phenomena. Moreover, AI-driven systems can optimize the allocation of resources, guiding astronomers and space agencies toward areas of the universe most likely to yield significant discoveries. Furthermore, AI platforms are integral in enhancing the sensitivity and accuracy of instruments used in cosmic observation, and fine-tuning parameters to filter out interference and extraneous noise. As space exploration ventures extend beyond our solar system, the implementation of AI promises to revolutionize the way we perceive and investigate the cosmos. Its ability to continually learn and adapt positions AI as an invaluable asset in the quest to unravel the mysteries of the universe. By combining human expertise with AI-driven technologies, cosmic search strategies stand poised to uncover unprecedented insights into the nature of celestial bodies, galactic structures, and potentially even extraterrestrial life. The integration of AI in cosmic search strategies heralds a new era of discovery, propelling humanity toward a deeper understanding of the cosmos and our place within it.

Expanding the Range: Beyond our Solar System

As we continue to push the boundaries of cosmic exploration, our pursuits extend far beyond the confines of our solar system. The quest to expand our reach into the cosmos has been a driving force behind groundbreaking advancements in spacefaring technologies and observational techniques. At the forefront of this expansion is the endeavor to identify and explore exoplanetary systems, each offering unique insights into the diversity and potential for life beyond our own celestial neighborhood.

The discovery of exoplanets orbiting distant stars has opened new frontiers in our understanding of planetary formation, habitability, and the potential prevalence of life-sustaining conditions throughout the universe. Utilizing advanced telescopic observations, including the remarkable capabilities of next-generation space observatories, we aim to scrutinize these distant worlds, deciphering their atmospheric compositions, surface features, and orbital dynamics. Such detailed examinations hold immense promise for uncovering clues about the prospects of extraterrestrial life and further elucidating the intricate tapestry of the cosmic ecosystem.

Moreover, as we venture into the uncharted realms beyond our solar system, astronomers and astrophysicists are deeply engaged in constructing sophisticated models and simulations to simulate the environments and conditions prevalent in these stellar systems. These simulations contribute significantly to our ability to interpret observational data, predict planetary characteristics, and delineate the potential habitability zones within these distant planetary systems. Through interdisciplinary collaborations and computational innovations, we strive to gain a comprehensive understanding of the diverse planetary architectures and climatic phenomena that shape the astronomical landscapes sprawled across the interstellar vastness.

In parallel, efforts are underway to design and deploy advanced spacecraft missions dedicated to the exploration of exoplanets and their host stars. Robotic explorers equipped with state-of-the-art instrumentation are envisioned to embark on journeys spanning light-years, enabling close-up investigations of alien worlds and the celestial bodies that punctuate their starry realms. These missions not only serve as pioneering initiatives in space technology but also hold the potential to revolutionize our perceptions of planetary systems, enriching our insights into the cosmic panorama that extends far beyond the planetary boundaries we presently call home.

The relentless pursuit to expand the range of our cosmic explorations beyond our solar system encapsulates an awe-inspiring odyssey of human ingenuity, scientific curiosity, and unwavering determination. It propels us towards a future where the mysteries of distant worlds become unraveled, and the vistas of the cosmos beckon us with tantalizing prospects of discovery and revelation.

Interdisciplinary Approaches to Cosmic Investigation

Interdisciplinary collaboration has become increasingly critical in the realm of cosmic investigation. As humanity ventures beyond our solar system, the complex and multifaceted nature of cosmic exploration demands a convergence of various disciplines. In this quest to unravel the mysteries of the universe and potentially discover extraterrestrial life, the integration of diverse fields such as astrophysics, astrobiology, geology, biology, chemistry, computer science, and engineering is paramount. The synergy between these disciplines not only broadens our holistic understanding but also accelerates progress in cosmic research.

Astrophysics constitutes a fundamental pillar in cosmic investigation, providing insights into celestial bodies and their interactions. Through collaborative efforts with other scientific domains, astrophysical data can be analyzed alongside biological and geological knowledge to identify potential habitable exoplanets or peculiar cosmic phenomena indicative of extraterrestrial existence. Furthermore, the synthesis of biochemistry and exobiology offers unique perspectives on the plausibility of life outside Earth, fostering interdisciplinary discussions that refine our approach to cosmic exploration.

The involvement of computer science and artificial intelligence (AI) is indispensable in processing vast quantities of data obtained from sophisticated telescopes and planetary probes. By integrating AI algorithms, interdisciplinary teams can navigate through extensive datasets more effectively, discerning patterns that elude traditional analysis methods. Moreover, the fusion of advanced communication technologies and robotics with astrobiology and astrogeology enables the development of autonomous systems capable of exploring and investigating distant cosmic environments, exponentially expanding our exploratory reach.

In the pursuit of cosmic investigation, geologists contribute their expertise in analyzing extraterrestrial materials and structures, drawing parallels between terrestrial and extraterrestrial landscapes to infer the presence of ancient civilizations or remnants of cosmic activity. Chemistry provides vital support by facilitating the identification of biomarkers and chemical signatures that indicate the potential for life beyond Earth. Such interdisciplinary interactions not only enable comprehensive assessments of cosmic environments but also cultivate innovative methodologies for detecting and characterizing extraterrestrial phenomena.

As we continue to push the boundaries of cosmic exploration, harmonious collaboration among diverse scientific fields is indispensable. Interdisciplinary approaches not only enrich the depth of our inquiries but also foster innovation, driving the development of novel technologies and methodologies that propel humanity toward an unprecedented understanding of the cosmos.

Enhancing Deep Space Communication Systems

Advancements in deep space communication systems have long been a crucial focus of the scientific community, driven by the necessity to establish robust and reliable means of communication across vast cosmic distances. As we venture further into the uncharted realms of space, the inherent challenges posed by interstellar communication demand innovative solutions that transcend conventional technologies. The quest to enhance deep space communication systems encompasses a multifaceted approach involving cutting-edge developments in signal processing, antenna technologies, and data transmission protocols. To overcome the impediments presented by cosmic distances and signal degradation, pioneering research is being conducted to refine communication protocols and develop novel methods for transmitting and receiving data across astronomical distances. Additionally, the utilization of advanced beamforming techniques and adaptive modulation schemes holds immense promise in optimizing communication links within the deep space environment. Furthermore, the integration of artificial intelligence (AI) and machine learning algorithms has emerged as a transformative force in enhancing the efficiency and reliability of deep space communication systems. By leveraging AI technologies, scientists are striving to mitigate the effects of signal attenuation and distortion, thereby enabling more resilient and error-tolerant communication channels. Moreover, the deployment of autonomous communication networks equipped with cognitive capabilities empowers these systems to adapt dynamically to the evolving conditions of deep space, ensuring seamless and uninterrupted data exchange. Collaborative efforts between space agencies and telecommunications experts are pivotal in driving forward the frontier of deep space communication, fostering the development of next-generation communication architectures designed to withstand the rigors of space exploration. The evolution of quantum communication principles presents a paradigm-shifting prospect for revolutionizing deep-space communication. Quantum entanglement-based communication holds the potential to enable instantaneous data transmission over vast cosmic distances, circumventing the constraints imposed by light-speed limitations. While still in its infancy, ongoing research endeavors aimed at harnessing quantum phenomena for interstellar communication herald a new era in the domain of cosmic information exchange. In tandem with technological advancements, ethical considerations surrounding deep space communication warrant meticulous examination. As humanity extends its reach into the cosmos, it becomes imperative to establish ethical frameworks governing communication with potential extraterrestrial civilizations, acknowledging the profound implications of interstellar interactions. Amidst the intricate tapestry of celestial communication, the pursuit of enhancing deep space communication systems stands as a cornerstone of our quest to unravel the enigmas of the cosmos.

Role of Quantum Computing in Space Research

Quantum computing stands at the forefront of technological advancements, poised to revolutionize space research and exploration. Traditional computers are becoming increasingly limited in their ability to handle the complex computational demands of space missions, from processing vast volumes of data to simulating intricate celestial phenomena. In contrast, quantum computing harnesses the principles of quantum mechanics to perform calculations at an unprecedented scale, offering a transformative paradigm for space-based endeavors.

The distinguishing feature of quantum computing lies in its utilization of quantum bits or qubits, which can represent multiple states simultaneously, enabling parallel processing on an exponential scale. This characteristic empowers quantum computers to solve problems that are currently insurmountable with classical computing methods. In the context of space exploration, this opens up a realm of possibilities, from optimizing orbital trajectories and navigation to modeling cosmic events with unparalleled accuracy.

One of the most compelling applications of quantum computing in space research is the advancement of secure communication protocols through quantum cryptography. Quantum entanglement enables the creation of unbreakable cryptographic keys, offering impregnable security for sensitive transmissions between ground control and space probes, protecting against potential interception by hostile entities.

Moreover, quantum computing holds the potential to accelerate the analysis of complex astrophysical data gathered from telescopes and space probes. By leveraging quantum algorithms, researchers can rapidly process immense datasets, unveiling new insights into celestial phenomena and expediting the discovery of exoplanets, gravitational waves, and other cosmic phenomena.

As space agencies set their sights on manned missions to distant planets and celestial bodies, quantum computing presents a critical tool for optimizing mission planning, resource management, and survivability assessments. The ability of quantum computers to tackle multifaceted optimization challenges could redefine mission architecture, fuel consumption analyses, and trajectory optimization, ultimately paving the way for more efficient and ambitious interstellar travel.

Looking ahead, the integration of quantum computing in space research underscores the need for interdisciplinary collaboration between astrophysicists, computer scientists, and quantum engineers. As quantum computing continues to mature, the space exploration community must embrace this disruptive technology, fostering a seamless fusion of quantum computing expertise with domain-specific astrophysical knowledge.

Legal and Ethical Considerations in Cosmic Exploration

The expansion of cosmic exploration also brings forth a myriad of legal and ethical considerations that necessitate careful contemplation. As humanity ventures further into the cosmos, it is essential to establish a robust framework for governing activities beyond Earth's boundaries. One of the key legal considerations involves defining ownership and governance of celestial bodies, resources, and potential discoveries. The formulation of international laws and agreements regarding space exploration, mining, and colonization becomes paramount as various entities aim to assert their presence in the vastness of space. This presents a complex landscape where international cooperation and diplomacy are essential to navigate potential disputes and conflicts. Ethics play an equally critical role in cosmic exploration, prompting reflection on the impact of human activities on extraterrestrial environments and any potential forms of life encountered. Additionally, consideration must be given to the preservation of cultural and historical sites beyond Earth, should such artifacts be discovered. Furthermore, the implications of communicating with extraterrestrial intelligence raise significant ethical questions, urging deliberation on the potential consequences and responsibilities associated with such contact. Questions about how human values and norms may influence interactions with alien civilizations also demand careful examination. Ethical guidelines and protocols need to be crafted to guide our interactions and minimize any detrimental effects on alien cultures or ecosystems. These considerations underscore the necessity for a balanced approach that harmonizes scientific pursuits with respect for the unknown and ensures the sustainable and responsible advancement of cosmic exploration. As advancements in technology and capabilities enable humankind to extend its reach into the universe, addressing legal and ethical aspects is pivotal in promoting a harmonious and principled expansion into the cosmos.

Collaborative International Space Missions

In an era of unprecedented technological advancement and global connectivity, the exploration of outer space has evolved into a collaborative endeavor that transcends national boundaries. The realization that the mysteries of the cosmos can only be fully understood through cooperation has spurred the development of numerous collaborative international space missions. These ambitious projects represent the epitome of scientific collaboration, as they bring together the expertise, resources, and ingenuity of multiple nations toward achieving common cosmic goals. Collaborative international space missions serve as a testament to humanity's shared curiosity about the universe and our collective determination to unlock its secrets. Through partnerships such as the International Space Station (ISS), the European Space Agency's (ESA) joint ventures, and collaborative efforts between NASA and other space agencies, countries are working hand in hand to expand our knowledge of the cosmos. These missions enable scientists, engineers, and astronauts from different nations to pool their talents and resources, fostering an environment of mutual learning and innovation. Furthermore, collaborative international space missions not only advance our understanding of the universe but also promote diplomatic ties and goodwill among participating countries. By engaging in these joint ventures, nations demonstrate their commitment to peaceful cooperation in the pursuit of scientific discovery. Moreover, the pooling of financial and human resources for space exploration not only makes missions more feasible but also helps distribute the costs and risks associated with venturing beyond Earth's atmosphere. As we look to the future, the continued success and expansion of collaborative international space missions will depend on the willingness of nations to prioritize collective advancement over individual interests. Governments, space agencies, and international organizations must maintain open lines of communication and establish transparent frameworks for sharing knowledge, technology, and resources. This collaborative spirit will not only propel us further into the cosmos but also serve as a beacon of hope for global cooperation and progress. As we explore new frontiers together, each mission becomes a symbol of our shared aspirations and a reminder of the boundless potential that awaits us beyond the stars.

The Future of Space-Based Telescopes and Observatories

Space-based telescopes and observatories have historically played a pivotal role in expanding our understanding of the cosmos. Looking toward the future, technological advancements in space-based observation are poised to revolutionize our capacity to explore the universe. The advent of sophisticated instrumentation, enhanced imaging technologies, and innovative data processing methodologies represents a significant leap forward in the capabilities of these instruments.

The deployment of next-generation space-based telescopes promises to provide unparalleled insights into celestial phenomena, from exoplanet discovery to the elucidation of the fundamental properties of dark matter and dark energy. These advanced telescopes and observatories are anticipated to possess unprecedented sensitivity and resolution, enabling the detailed study of distant galaxies, star formations, and cosmic phenomena previously beyond the reach of traditional ground-based observatories.

Moreover, the integration of state-of-the-art spectroscopic instruments and advanced adaptive optics systems will enable astronomers and astrophysicists to probe the composition and atmospheres of exoplanets with an unprecedented level of precision. This holds the potential to unlock profound insights into the existence of habitable worlds beyond our solar system and assess their potential for harboring life.

Furthermore, the future of space-based telescopes and observatories entails the utilization of cutting-edge propulsion systems and modular design architectures, which will facilitate in-orbit repair and reconfiguration, prolonging the operational lifespan of these instruments. This approach not only enhances the resilience and sustainability of space-based observatories but also ensures continuity in the collection of invaluable scientific data, minimizing interruptions due to technical malfunctions or component degradation.

Additionally, advancements in communication technologies and data transmission protocols will further streamline the real-time transfer of astronomical data from these space-based platforms to terrestrial research facilities. The seamless integration of these data streams with advanced machine learning algorithms and artificial intelligence-driven analytics will empower researchers to extract deeper insights from the unprecedented volume and complexity of extraterrestrial observations, elevating the frontiers of astrophysical research to new heights.

As we venture into this promising era of space-based exploration, international collaboration, and interdisciplinary partnerships will be instrumental in fostering the sustainable development and operation of these sophisticated instruments. By capitalizing on the expertise and resources of diverse nations and research organizations, we can collectively propel the frontiers of astronomical discovery, unveiling the mysteries of the universe and enriching our collective understanding of the cosmos.

Evaluating and Mitigating Cosmic Exploration Risks

Exploring the cosmos is an endeavor that presents unique challenges and potential risks. As humanity ventures further into space, it becomes imperative to meticulously evaluate and effectively mitigate these cosmic exploration risks. The remoteness and inhospitable nature of outer space introduce a multitude of hazards that must be addressed with careful consideration.

One of the primary risks in cosmic exploration is the impact of prolonged radiation exposure. Beyond the protective shield of the Earth's atmosphere, astronauts and equipment are exposed to high-energy particles from cosmic rays and solar radiation. Mitigating this risk requires advanced shielding technologies and pharmaceutical interventions to minimize the long-term effects of radiation on human health.

Additionally, the extended duration of space missions poses significant physiological and psychological challenges for astronauts. Addressing the potential health impacts of prolonged weightlessness, muscle atrophy, and bone density loss is crucial in ensuring the well-being of space travelers. Moreover, the isolation and confinement experienced during extended missions necessitate comprehensive mental health support systems and interpersonal dynamics management strategies.

Furthermore, the unpredictable and hostile environment of space introduces the inherent risk of space debris and micrometeoroid impacts on spacecraft. Implementing robust shielding and detection systems, as well as developing debris mitigation strategies, are essential in safeguarding mission-critical equipment and personnel from potential collisions.

The reliance on advanced life support systems and complex engineering technologies within confined spacecraft also introduces the vulnerability of critical system failures. Developing redundant systems, rigorous testing protocols, and fail-safe mechanisms are essential in mitigating the risks associated with technical malfunctions and ensuring the sustainability of space missions.

Moreover, the intricate nature of deep space exploration necessitates meticulous planning for extraterrestrial contingency situations, such as medical emergencies, communication breakdowns, or unanticipated environmental threats. Establishing comprehensive emergency response protocols and equipping personnel with extensive training in problem-solving and adaptability is paramount in addressing unforeseen challenges.

Finally, the ethical considerations of cosmic exploration risks cannot be overlooked, especially concerning the potential impact of human activities on extraterrestrial environments and the implications of microbial contamination. Developing stringent planetary protection measures and adhering to international space treaties are essential facets of responsible cosmic exploration.

In conclusion, evaluating and mitigating cosmic exploration risks demands a multidisciplinary approach, integrating advancements in technology, medicine, psychology, engineering, and ethics. By proactively addressing these challenges, humanity can ensure the safe and sustainable expansion of our presence in the cosmos while advancing our understanding of the universe.

Chapter Eighteen

Alien Intrigue: Unveiling Otherworldly Mysteries

Philosophical Perspectives

Humanity's fascination with the possibility of extraterrestrial life has long been intertwined with philosophical inquiry. The prospect of encounters with beings from other worlds raises profound questions about our place in the cosmos and the nature of existence itself. Throughout history, philosophers have grappled with the implications of alien life, exploring a diverse array of perspectives that illuminate the human quest for knowledge and understanding. From ancient civilizations pondering the existence of celestial beings to the Enlightenment era's speculation on inhabited planets, the concept of extraterrestrial life has sparked contemplation across cultures and epochs. Philosophers have probed the implications of potential contact with intelligent extraterrestrial civilizations, delving into profound ethical, existential, and epistemological considerations. In this chapter, we embark on a journey through the annals of philosophical thought, tracing the evolution of ideas surrounding extraterrestrial existence and contemplating the profound ramifications of such a cosmic discovery. Drawing upon a tapestry of philosophical traditions, we will examine how different cultural, religious, and intellectual frameworks have shaped humanity's conceptualization of alien life forms and their potential impact on our collective worldview. As we navigate this boundless realm of thought, we delve into the rich tapestry of philosophical perspectives, seeking to unravel the complexities of human consciousness in light of the vast unknowns of the universe.

Historical Overview of Philosophical Thought on Extraterrestrial Life

Throughout history, the concept of extraterrestrial life has intrigued and inspired philosophers across various cultures and epochs. In ancient Greece, the idea of alien beings beyond Earth was contemplated by thinkers such as Anaximander and Epicurus. Anaximander proposed that other worlds existed, while Epicurus speculated on the plurality of worlds and the possibility of diverse forms of life. These discussions laid a foundation for the philosophical exploration of extraterrestrial existence. Moving forward in time, medieval and Renaissance philosophers contemplated the nature of celestial bodies and the potential for life beyond our planet. Visionaries like Giordano Bruno dared to challenge the geocentric worldview with radical cosmological ideas, advocating for the existence of infinite inhabited worlds. The Copernican Revolution further reshaped philosophical perspectives, opening up the possibility of Earth being just one of many inhabited planets in the cosmos. The Enlightenment period saw a surge of interest in the plurality of worlds and the philosophical implications of potential contact with extraterrestrial beings. Figures like Immanuel Kant and David Hume delved into the philosophical implications of a universe teeming with inhabited worlds, grappling with questions of cosmic significance and the limits of human knowledge. In the modern era, technological advancements and astronomical discoveries have contributed to renewed philosophical explorations of extraterrestrial life. The advent of space exploration, astrobiology, and the search for exoplanets has reignited debates surrounding the philosophical, ethical, and existential aspects of potential extraterrestrial encounters. With each era building upon the intellectual legacy of its predecessors, the historical overview of philosophical thought on extraterrestrial life reveals a rich tapestry of ideas, controversies, and speculations that continue to shape our understanding of the cosmos and our place within it.

Epistemological Challenges in Understanding Alien Intelligence

The search for extraterrestrial intelligence poses profound epistemological challenges, forcing us to question the very foundations of our understanding of intelligence and cognition. As we embark on this intellectual journey, we are confronted with the limitations of our own cognitive frameworks and the inadequacy of our terrestrial-centric perspectives. One of the primary epistemological challenges lies in defining what constitutes intelligence beyond the scope of human experience. Our current methods of determining and measuring intelligence may be ill-equipped to grasp the diverse forms of cognitive processes that could exist among alien species. Furthermore, the reliance on familiar patterns and paradigms derived from terrestrial life forms may hinder our ability to recognize and interpret alien intelligence. The very concept of 'intelligence' itself becomes a point of contention, as it is invariably shaped by our human biases and cultural constructs. Another fundamental challenge arises in discerning communication and symbolic systems that may be radically different from our own. The assumptions underlying our tools for deciphering languages and symbols may prove inadequate when faced with the complexity of extraterrestrial communication systems. This not only pertains to linguistic communication but also includes non-verbal forms of expression, such as gestural, visual, or olfactory cues that may be entirely alien to us. Moreover, the epistemological obstacles extend to the very nature of understanding an intelligence that might operate on principles fundamentally divergent from those underpinning human cognition. These challenges prompt us to question the universality and relativity of cognitive norms, challenging the very essence of what it means to understand and be understood. In navigating these epistemological frontiers, interdisciplinary collaboration across fields such as linguistics, anthropology, cognitive science, and information theory becomes essential. It requires a reimagining of our established epistemic frameworks and a willingness to embrace radical new perspectives. Ultimately, grappling with the epistemological challenges of understanding alien intelligence presents an opportunity to transcend the boundaries of human cognition and glimpse the vast tapestry of possibilities that await discovery in the cosmic realm.

Ontological Considerations and the Concept of 'Otherness'

Ontological considerations regarding the concept of 'otherness' are at the heart of philosophical discourse on extraterrestrial existence. This concept delves into the fundamental nature of beings or entities that are distinct from ourselves, in this case, extraterrestrial intelligence. It prompts an exploration of the boundaries of identity and the implications of encountering life forms that challenge our understanding of existence. Philosophical debates often revolve around the potential impact of interacting with beings whose very essence transcends the familiar paradigms of human experience. The concept of 'otherness' forces us to reevaluate our ontological framework, as it compels us to confront the profound ramifications of encountering life forms beyond Earth. This leads to a reconsideration of our place in the universe, as the recognition of alien otherness challenges anthropocentric assumptions about the centrality of human existence. Additionally, contemplating the concept of 'otherness' raises questions about the nature of consciousness, intelligence, and agency beyond terrestrial limitations. These considerations prompt us to reassess the boundaries of personhood, cognition, and ethical considerations in relation to non-human entities. The ontological quandaries surrounding the concept of 'otherness' in the context of extraterrestrial existence thus necessitate an expanded philosophical inquiry into the nature of being, identity, and the foundations of interspecies interactions. This exploration of 'otherness' lays the groundwork for redefining our existential outlook and reshaping our understanding of the universe in light of potential encounters with extraterrestrial life.

Ethics of First Contact: Protocols and Responsibilities

As humanity delves into the realm of potential extraterrestrial contact, a crucial aspect that demands attention is the ethical considerations surrounding first contact protocols and the responsibilities associated with such an event. The implications of encountering alien life forms extend far beyond the realms of science and technology; they seep into the very fabric of human ethics, societal structures, and global governance. In contemplating the ethics of first contact, it becomes imperative to establish a framework imbued with prudence, respect, and cautious optimism. This framework must prioritize the preservation of mutual dignity and the avoidance of harm, whether intentional or inadvertent, to both humanity and any extraterrestrial civilizations. The establishment of universally acknowledged guidelines, informed by interdisciplinary perspectives encompassing philosophy, ethics, law, and diplomacy, is indispensable. Such protocols should provide a roadmap for respectful, peaceful, and mutually beneficial engagement with potential extraterrestrial entities. Central to this ethical framework is the acknowledgment of potential disparities in technological, cultural, and cognitive capabilities between humanity and alien civilizations. The onus lies on humanity to approach first contact with humility, empathy, and a keen awareness of our own inherent biases. Respect for the autonomy and self-determination of any discovered extraterrestrial life forms should remain at the core of our interactions. Furthermore, as stewards of Earth and representatives of our species, humans bear the responsibility of ensuring that any contact with extraterrestrial intelligence contributes positively to the advancement of knowledge, understanding, and cooperation across civilizations. Aiding in building a foundation of trust and cooperation, ethical protocols must also address the complexities of communication with entities possessing potentially incomprehensible modes of expression and cognition. Sensitivity to nuances of cross-cultural communication, coupled with the deployment of rigorous ethical review processes, will be vital in minimizing misunderstandings and conflicts during initial exchanges. The establishment of a transparent, inclusive, and globally agreed-upon system for assessing the potential impacts and risks of first contact is pivotal. Transparency fosters trust and ensures that decisions are made collectively and responsibly, reflecting the diverse values and interests of humanity. Lastly, ethical frameworks for first contact must grapple with the intricacies of long-term coexistence and collaboration with extraterrestrial beings, if such interactions are to unfold. This encompasses the development of principles guiding cooperative endeavors, the sharing of knowledge, and the ethically sound utilization of resources. By proactively addressing these profound ethical dimensions, humanity can strive to elevate the prospect of first contact from mere scientific curiosity to a beacon of wisdom, compassion, and unity amidst the mysteries of the cosmos.

Anthropocentrism and Its Limits in a Universe with Alien Life

Anthropocentrism, the notion that human beings are the central or most significant entities in the universe, has been deeply ingrained in the philosophical, scientific, and cultural narratives of humanity. However, the potential discovery of extraterrestrial life challenges this deeply rooted anthropocentric perspective. The existence of intelligent alien civilizations would necessitate a dramatic shift in our worldview, forcing us to reevaluate the place of humanity in the broader cosmic community.

Historically, anthropocentrism has influenced human attitudes towards the natural world, often leading to exploitation and disregard for non-human life forms. The assumption of human uniqueness and superiority has shaped our ethical frameworks, social structures, and scientific endeavors. Confronting the prospect of coexisting with technologically advanced extraterrestrial species demands a critical examination of the limitations of anthropocentrism.

The realization of extraterrestrial intelligence prompts the need for an expanded moral consideration towards non-human beings. It challenges us to transcend the boundaries of human-centric values and ethics, fostering a more inclusive and expansive ethical stance that acknowledges the potential diversity of moral perspectives in a cosmological context. This recognition invites us to reconsider our environmental responsibilities not only to our planet but also to potential alien ecosystems and their inhabitants.

Moreover, the emergence of alien civilizations would prompt profound philosophical reflections on the nature of consciousness, intelligence, and personhood. Anthropocentrism asserts the uniqueness of human cognitive abilities and the centrality of human experiences, yet the encounter with alternative forms of intelligence would compel a reevaluation of these assumptions. Exploring the cognitive capacities and societal structures of extraterrestrial species would enrich our understanding of intelligence, challenging the exclusivity of human cognition and its associated definitions.

The limits of anthropocentrism become starkly evident in the face of potential contact with advanced alien societies. Such encounters could revolutionize our conceptions of identity, purpose, and meaning, inviting us to engage in a more expansive dialogue about humanity's place within the cosmic tapestry. While anthropocentrism has provided a basic framework for understanding our position in the universe, the prospect of interacting with alien life underscores the imperative for a broader, more pluralistic cosmic perspective, one that transcends the inherent biases of human-centric narratives.

Existential Risks and Benefits of Extraterrestrial Contact

The prospect of extraterrestrial contact presents humankind with unprecedented existential risks and potential benefits. As we ponder the implications of such encounters, it is essential to consider the multifaceted nature of these potential consequences. Firstly, from a risk perspective, the uncharted territory of engaging with alien civilizations brings forth the possibility of unknown threats to humanity. These include both biological and ideological hazards, as well as the potential for conflict arising from differences in values, ethics, and objectives. The very act of making contact could inadvertently expose our species to unfamiliar diseases or environmental factors, posing significant risks to terrestrial ecosystems and human life. Moreover, the dissemination of advanced alien technologies may disrupt existing power dynamics and socio-economic structures on Earth, leading to global instability. Conversely, the benefits of extraterrestrial contact are equally profound. Such an encounter could offer immeasurable advancements in scientific knowledge, fostering breakthroughs in fields such as physics, biology, and sociology. The exchange of ideas between civilizations with vastly divergent histories and perspectives could enrich our collective understanding of the universe and our place within it. Additionally, the potential for collaboration in addressing shared cosmic challenges, such as celestial threats or interstellar travel, holds promise for the collective progress of sentient beings across the cosmos. Furthermore, the realization that we are not alone in the universe may prompt a reevaluation of human identity and purpose, transcending terrestrial divides and conflicts toward a more unified planetary civilization. Embracing the unknown while diligently preparing for its potential impacts is paramount in crafting a holistic approach to the existential risks and benefits entwined with the possibility of extraterrestrial contact.

Philosophical Theories of Mind Applied to Alien Intelligence

The philosophical exploration of alien intelligence necessitates a deep dive into various theories of mind. In contemplating the nature of extraterrestrial cognition, scholars and thinkers have drawn upon foundational philosophical concepts to form a framework for understanding potential forms of alien consciousness. One fundamental aspect of this inquiry is the consideration of panpsychism, the idea that consciousness is an inherent property of the physical universe. Proponents of this view posit that all entities, even at the cosmic level, possess some form of conscious experience. By extrapolating panpsychist principles to the prospect of extraterrestrial beings, we are prompted to question the nature of their cognitive processes and the scope of their awareness within the vast cosmos. Expanding on this, the philosophical examination of emergent properties in complex systems sheds light on the potential manifestation of intelligence in distinct biological or technological forms. This aligns with the conjecture that alien minds may exhibit modes of intelligence that diverge significantly from human cognitive paradigms. Moreover, parallels can be drawn between phenomenological approaches to consciousness and the hypothetical mental states of extraterrestrial entities. The project of phenomenology, which seeks to elucidate the structures of subjective experience, offers a window into how alien minds might perceive and interpret reality in ways unfathomable to us. Contemplating the transcendence of cultural and linguistic barriers, the philosophy of language also plays a crucial role in conceptualizing the cognitive capacities of alien intelligences. Through conceptual challenges such as qualia and the nature of representation, philosophers have grappled with the notion of communicating with entities possessing radically different cognitive architectures. Furthermore, the philosophical exploration of artificial intelligence provides valuable insights into the potential diversity of intelligent systems, shedding light on the plausibility of encountering non-biological forms of consciousness among extraterrestrial civilizations. These interdisciplinary inquiries into the philosophical theories of mind not only enrich our theoretical understanding of alien intelligence but also inform our perspectives on the profound diversity of conscious experience across the cosmos.

Implications for Human Identity and Unity

The potential discovery of extraterrestrial life raises profound questions about the impact on human identity and unity. In contemplating the existence of intelligent beings beyond Earth, humans are confronted with reevaluating their place in the universe. The concept of 'self' and 'other' takes on new dimensions as humanity grapples with the idea of coexisting alongside alien civilizations. This reevaluation challenges traditional notions of human exceptionalism and forces a reassessment of how human society defines itself in relation to a broader cosmic community. Moreover, the prospect of interacting with extraterrestrial beings presents an opportunity to foster global unity and cooperation. Such an encounter may serve as a catalyst for transcending earthly conflicts and divisions, prompting a collective redefinition of what it means to be human in a cosmic context. The shared experience of initiating contact with alien intelligence could potentially unite disparate factions of humanity under a common purpose, stimulating a renaissance of global collaboration and understanding. At the same time, the realization that humans are not alone in the cosmos has the potential to instill a profound sense of interconnectedness and empathy among inhabitants of Earth. This expanded awareness of our place in the universe may engender a greater sense of responsibility towards not only our planet but also towards potential extraterrestrial neighbors. Furthermore, the shift in perspective induced by the recognition of other intelligent beings has the capacity to dissolve artificial barriers that have historically divided humankind. Through the lens of an extended cosmic community, differences in race, nationality, and ideology may pale in comparison to the shared universal experience of being part of an inhabited cosmos. As such, the implications for human identity and unity in a scenario involving the confirmation of extraterrestrial life are multifaceted and offer a profound opportunity for reflection and transformation.

Future Directions in the Philosophy of Extraterrestrial Existence

As humankind embarks on an era of heightened exploration and potential discovery of extraterrestrial life, the field of philosophy faces critical junctures in its discourse surrounding the implications of such profound encounters. Looking to the future, several key areas warrant sustained attention and scholarly inquiry. First and foremost, the ethical and moral considerations associated with engaging with extraterrestrial beings demand rigorous exploration. Philosophers must delve into questions of universal rights, responsibilities towards non-human entities, and the establishment of ethical frameworks that can guide interstellar interactions. Moreover, as advancements in astrobiology and space exploration technology continue to unfold, it becomes imperative for philosophers to actively engage in shaping interdisciplinary dialogues and contributing to policy-making efforts that promote responsible research and exploration endeavors. This multidisciplinary approach will safeguard against hasty or reckless actions that could have far-reaching consequences for human society and any potential extraterrestrial civilization. Additionally, the ontological and epistemological dimensions of encounter scenarios remain fertile ground for philosophical investigation. The very nature of how we define 'being' and 'intelligence' will undergo a profound reevaluation in light of potential contact with extraterrestrial life forms. Thought experiments and theoretical explorations within this domain will help elucidate the fundamental aspects of our understanding of existence and cognition. Furthermore, the emergence of disruptive technologies such as quantum computing and advancements in artificial intelligence will likely reshape philosophical discussions about the nature of consciousness, agency, and communication across cosmic distances. Finally, the prospect of establishing open channels for meaningful dialogue and collaboration with any discovered extraterrestrial civilizations presents a compelling area for exploration. Philosophers are poised to contribute valuable insights towards designing communication protocols, deciphering potential extraterrestrial messages, and fostering mutual understanding rooted in cultural sensitivity. It is incumbent upon the philosophical community to proactively engage with these and other pertinent issues to guide humanity towards a future where the profound implications of extraterrestrial existence are met with intellectual rigor, ethical clarity, and a steadfast commitment to the betterment of both terrestrial and cosmic societies.

Chapter Nineteen

Bibliographic References

A. Allwood, D. Beaty, D. Bass, C. Conley, G. Kminek, M. Race, S. Vance and F. Westall. "Conference summary: life detection in extraterrestrial samples.." *Astrobiology*, 13 2 (2013): 203-16 . https://doi.org/10.1089/a st.2012.0931.

A. Burton, J. Stern, J. Elsila, D. Glavin and J. Dworkin. "Understanding prebiotic chemistry through the analysis of extraterrestrial amino acids and nucleobases in meteorites.." *Chemical Society reviews*, 41 16 (2012): 5459-72 . https://doi.org/10.1039/c2cs35109a.

A. Frebel and J. Norris. "Near-Field Cosmology with Extremely Metal-Poor Stars." *Annual Review of Astronomy and Astrophysics*, 53 (2015): 631-688. https://doi.org/10.1146/annurev-astro-082214-122423.

A. Heavens, B. Panter, R. Jimenez and J. Dunlop. "The star-formation history of the Universe from the stellar populations of nearby galaxies." *Nature*, 428 (2004): 625-627. https://doi.org/10.1038/nature02474.

Avi Loeb. *Extraterrestrial: The First Sign of Intelligent Life beyond Eart*. Houghton Mifflin Harcourt, 2021.

Barlow, Martin T. "Galactic Exploration by Directed Self-Replicating Probes, and Its Implications for the Fermi Paradox." *International Journal of Astrobiology* 12, no. 1 (October 3, 2012): 63–68. https://doi.org/10.1017/s1473550412000419.

Blau, J., and Nelson F. Adkins. "Philip Freneau and the Cosmic Enigma," 1949.

B. Robertson, R. Ellis, J. Dunlop, R. McLure and D. Stark. "Early star-forming galaxies and the reionization of the Universe." *Nature*, 468 (2010): 49-55. https://doi.org/10.1038/nature09527.

B. Schmitz. "Extraterrestrial spinels and the astronomical perspective on Earth's geological record and evolution of life." *Chemie der Erde*, 73 (2013): 117-145. https://doi.org/10.1016/J.CHEMER.2013.04.002.

Catling, David. *Astrobiology : A Very Short Introduction*. Oxford, United Kingdom: Oxford University Press, 2013.

Cockell, Charles. *Astrobiology : Understanding Life in the Universe*. Chichester, West Sussex, Uk: Wiley Blackwell, 2015.

"Cosmic Archaeology: Searching for Evidence of Past Technological Civilizations", Scientific American (2018).

Cowie, A., and Nelson F. Adkins. "Philip Freneau and the Cosmic Enigma: The Religious and Philosophical Speculations of an American Poet." American Literature 23 (1951): 144.

Crawford, Ian. *Expanding Worldviews: Astrobiology, Big History and Cosmic Perspectives*. Springer Nature, 2021.

Crowther, Thomas. "Bridging Gaps through Light : An Archaeological Exploration of Light and Dark in the Atlantic Scottish Iron Age," 2014.

C. Sagan, W. R. Thompson, R. Carlson, D. Gurnett and C. Hord. "A search for life on Earth from the Galileo spacecraft." *Nature*, 365 (1993): 715-721. https://doi.org/10.1038/365715A0.

Dartnell, Lewis. *Life in the Universe*. Simon and Schuster, 2012.

G. D. Silva, K. Freeman, J. Bland-Hawthorn, S. Martell, E. W. Boer, M. Asplund, S. Keller, S. Sharma, D. Zucker, T. Zwitter, B. Anguiano, C. Bacigalupo, D. Bayliss, M. Beavis, M. Bergemann, S. Campbell, R. Cannon, D. Carollo, L. Casagrande, A. Casey, G. Costa, V. D'Orazi, A. Dotter, L. Duong, A. Heger, M. Ireland, P. Kafle, J. Kos, J. Lattanzio, G. Lewis, J. Lin, K. Lind, U. Munari, D. Nataf, S. O'Toole, Q. Parker, W. Reid, K. Schlesinger, A. Sheinis, J. Simpson, D. Stello, Y. Ting, G. Traven, F. Watson, R. Wittenmyer, D. Yong and M. Žerjal. "The GALAH survey: scientific motivation." *Monthly Notices of the Royal Astronomical Society*, 449 (2015): 2604-2617. https://doi.org/10.1093 /mnras/stv327.

Hauman, Nicholas. "Heavenly Bodies: Scientific Racism and the Taxonomy of Extraterrestrial Life." Borderlands Journal 20 (2021): 124–54.

I. Crawford. "The astrobiological case for renewed robotic and human exploration of the Moon." *International Journal of Astrobiology*, 5 (2006): 191 - 197. https://doi.org/10.1017/S1473550406002990.

Jan Martel, David J. Young, Hsin-Hsin Peng, Cheng-Yeu Wu and John D. Young. "Biomimetic Properties of Minerals and the Search for Life in the Martian Meteorite ALH84001." *Annual Review of Earth and Planetary Sciences*, 40 (2012): 167-193. https://doi.org/10.1146/ANN UREV-EARTH-042711-105401.

Jason T. Wright, Jacob Haqq-Misra, Adam Frank, Ravi Kopparapu, Manasvi Lingam, Sofia Z. Sheikh."The Case for Technosignatures: Why They

May Be Abundant, Long-lived, Highly Detectable, and Unambiguous." https://doi.org/10.3847/2041-8213/ac5824

J. Campbell. "Archaeology and direct imaging of exoplanets." *Proceedings of the International Astronomical Union*, 1 (2005): 247 - 250. https://doi.org/10.1017/S1743921306009392.

Jin, Mingyu, Beichen Wang, Zhaoqian Xue, Suiyuan Zhu, Wenyue Hua, Hua Tang, Kai Mei, Mengnan Du, and Yongfeng Zhang. "What If LLMs Have Different World Views: Simulating Alien Civilizations with LLM-Based Agents." arXiv.Org abs/2402.13184 (2024).

Jurašeková, Z., G. Fabriciová, L. F. Silveira, Yuong-Nam Lee, Jaroslav M. Gutak, M. M. Ataabadi, and M. Kundrát. "Raman Spectra and Ancient Life: Vibrational ID Profiles of Fossilized (Bone) Tissues." International Journal of Molecular Sciences 23 (2022).

J. Wright. "Prior indigenous technological species." *International Journal of Astrobiology*, 17 (2017): 96 - 100. https://doi.org/10.1017/S1473550417000143.

Martin, A. "The Plurality of Worlds: A Synergistic Approach to the Question of the Search for Extraterrestrial Intelligence (SETI)." Futures 17 (1985): 618–31.

———. *Origins*. Basic Books, 2019.

Davies, Paul. *The Eerie Silence*. HarperCollins, 2010.

Dick, Steven J, Mark L Lupisella, and Aeronautics and. *Cosmos & Culture: Cultural Evolution in a Cosmic Context*. Government Printing Office, 2012.

Goldsmith, Donald, and Tobias C Owen. *The Search for Life in the Universe*. Sausalito, Calif.: University Science Books, 2002.

Juan Francisco Salazar, and Alice Gorman. *The Routledge Handbook of Social Studies of Outer Space*. Taylor & Francis, 2023.

Kevin Peter Hand. *Alien Oceans : The Search for Life in the Depths of Space*. Princeton, New Jersey: Princeton University Press, 2021.

Lincoln, Don. *Alien Universe*. JHU Press, 2013.

Lingam, Manasvi, and Abraham Loeb. *Life in the Cosmos : From Biosignatures to Technosignatures*. Cambridge, Massachusetts: Harvard University Press, 2021.

Louis Neal Irwin, Dirk Schulze-Makuch, and Springerlink (Online Service. *Cosmic Biology : How Life Could Evolve on Other Worlds*. New York, Ny: Praxis, 2011.

Mee, Nicholas. *The Cosmic Mystery Tour*. Oxford University Press, 2019.

Oliver, B.M. "SETI [Search for Extra-Terrestrial Intelligence]." *IEEE Potentials* 13, no. 3 (August 1994): 51–54. https://doi.org/10.1109/45.310941.

Osmanov, Z. "Are We Visible to Advanced Alien Civilizations?" Acta Astronautica, 2023.

P C W Davies. *The Eerie Silence : Are We Alone in the Universe?* London ; New York: Allen Lane, 2010.

Parcak, Sarah. *Archaeology from Space*. Henry Holt and Company, 2019.

Rahman, S. F. "The Enduring Enigma of the Cosmic Cold Spot." Physics World 33 (2020): 36–40.

R. Hoover. "Instruments and methods to search for extraterrestrial life." , 9606 (2015). https://doi.org/10.1117/12.2192893.

Robert A. Freitas and Francisco Valdes. "The search for extraterrestrial artifacts (SETA)." *Journal of the British Interplanetary Society* (1983). https://doi.org/10.1016/0094-5765(85)90031-1.

Rothery, David A, Iain Gilmour, and Mark A Sephton. *An Introduction to Astrobiology*. Cambridge University Press, 2018.

Rudolph, T. "Perhaps They Are Everywhere? Undetectable Distributed Quantum Computation and Communication for Alien Civilizations Can Be Established Using Thermal Light from Stars," 2021.

Schulze-Makuch, Dirk, Louis N Irwin, and Springerlink (Online Service. *Life in the Universe : Expectations and Constraints*. Cham: Springer International Publishing, 2018.

Simon Conway Morris. *From Extraterrestrials to Animal Minds*. Templeton Foundation Press, 2022.

Temple, Robert. *Sirius Mystery*. Century, 1998.

Vakoch, Douglas A. *Archaeology, Anthropology, and Interstellar Communication*. U. S. National Aeronautics & Space Administration, 2014.

V. Bromm. "Formation of the first stars." *Reports on Progress in Physics*, 76 (2005). https://doi.org/10.1088/0034-4885/76/11/112901.

Ward, Peter D, and Donald Brownlee. *Rare Earth*. Springer, 2007.

Webb, Stephen. *If the Universe Is Teeming with Aliens ... Where Is Everybody? : Seventy-Five Solutions to the Fermi Paradox and the Problem of Extraterrestrial Life*. 2nd ed. Cham, Switzerland: Springer, 2015.

Williams, Matt. "We Could Detect Alien Civilizations through Their Interstellar Quantum Communication," 2021.

Wright, Jason T., Shubham Kanodia, and Emily Lubar. "How Much SETI Has Been Done? Finding Needles in the N-Dimensional Cosmic Haystack." *The Astronomical Journal* 156, no. 6 (November 14, 2018): 260. https://doi.org/10.3847/1538-3881/aae099.

Wright, J.T. Milan M. Ćirković: The Great Silence: The Science and Philosophy of Fermi's Paradox. *Orig Life Evol Biosph* 48, 423–425 (2018). https://doi.org/10.1007/s11084-018-9568-3

Yashaswini Yadav. "The Hunt for Extraterrestrial Life: Exploring the Frontiers of Astrobiology." *International Journal For Multidisciplinary Research* (2023). https://doi.org/10.36948/ijfmr.2023.v05i05.6969.

www.ingramcontent.com/pod-product-compliance
Lightning Source LLC
LaVergne TN
LVHW052305200926
848481LV00002B/473